D0611187

Investors and Exploiters in Ecology and Economics

Principles and Applications

Strüngmann Forum Reports

Julia Lupp, series editor

The Ernst Strüngmann Forum is made possible through the generous support of the Ernst Strüngmann Foundation, inaugurated by Dr. Andreas and Dr. Thomas Strüngmann.

This Forum was supported by the
Deutsche Forschungsgemeinschaft
(The German Research Foundation)

Investors and Exploiters in Ecology and Economics

Principles and Applications

Edited by

Luc-Alain Giraldeau, Philipp Heeb, and Michael Kosfeld

Program Advisory Committee:

Luc-Alain Giraldeau, Philipp Heeb, Alex Kacelnik,
Michael Kosfeld, Julia Lupp, and Frédéric Thomas

The MIT Press

Cambridge, Massachusetts
London, England

© 2017 Massachusetts Institute of Technology and
the Frankfurt Institute for Advanced Studies

Series Editor: J. Lupp
Assistant Editor: M. Turner
Photographs: N. Miguletz
Lektorat: BerlinScienceWorks

The book was set in TimesNewRoman and Arial.
Printed and bound in the United States of America.

Library of Congress Cataloging-in-Publication Data is available.

ISBN: 978-0-262-03612-2

10 9 8 7 6 5 4 3 2 1

Contents

The Ernst Strüngmann Forum

Science is a highly specialized enterprise—one that enables areas of enquiry to be minutely pursued, establishes working paradigms and normative standards, and supports rigor in experimental research. Some issues, however, do not fall neatly into the purview of a single disciplinary field. Here, specialization can hinder conceptualization and limit the generation of potential problem-solving approaches. The Ernst Strüngmann Forum was created to explore these types of problems.

Founded on the tenets of scientific independence and the inquisitive nature of the human mind, the Ernst Strüngmann Forum is dedicated to the continual expansion of knowledge. Its activities promote interdisciplinary communication on high-priority issues encountered in basic science. Through its innovative communication process, the Ernst Strüngmann Forum provides a creative environment within which experts scrutinize high-priority issues from multiple vantage points.

This process begins with the identification of themes. By nature, a theme constitutes a problem area that transcends classic disciplinary boundaries, is of high-priority interest, and requires concentrated, multidisciplinary input to address the issues. Proposals are received from leading scientists active in their field and selected by an independent Scientific Advisory Board. Once approved, the Ernst Strüngmann Forum convenes a steering committee to refine the scientific parameters of the proposal and select participants. Approximately one year later, a central gathering, or Forum, is held to which circa forty experts are invited. Expansive discourse is employed to approach the problem. Often, this necessitates reexamining long-established ideas and relinquishing conventional perspectives. Yet when this is accomplished, new insights begin to emerge. As a final step, the resultant ideas and newly gained perspectives from the entire process are communicated to the scientific community for further consideration and implementation.

Preliminary discussion for this topic began in 2013, when Luc-Alain Giraldeau approached me, eager to develop a dialogue on the phenomenon of exploitation between multiple disciplines. Working together with Philipp Heeb, Luc-Alain submitted a proposal, and from January 9–11, 2015, the Program Advisory Committee (Luc-Alain Giraldeau, Philipp Heeb, Alex Kacelnik, Michael Kosfeld, Julia Lupp, and Frédéric Thomas) met to refine the scientific framework and select the participants to the Forum. From November 1–6, 2015, the invited group of anthropologists, biologists, economists, and public health scientists met in Frankfurt am Main to explore real-life cases and theoretical models of exploitation in humans and other organisms, and to work toward a common synthesis that would promote a unified framework under which implications for public health, natural resource use, and the design of institutions could be explored.

The extended discourse that ensued is captured in this volume, which is comprised of two types of contributions. Background information is provided on key aspects of the overall theme. These chapters, drafted before the Forum, were subsequently revised based on formal reviews and the input received at the Forum. Chapters 3, 6, 9, and 12 summarize the extensive discussions of the working groups. These chapters are not consensus documents nor are they proceedings; they transfer the essence of this multifaceted discourse, expose areas where opinions diverge, and highlight topics in need of future enquiry.

An endeavor of this kind creates its own unique group dynamics and puts demands on everyone who participates. Each invitee played an active role, and for their efforts, I am grateful to all. A special word of thanks goes to the Program Advisory Committee, to the authors and reviewers of the background papers, as well as to the moderators of the individual working groups (Alex Kacelnik, Bill Sutherland, Fred Thomas, and Michael Kosfeld). The rapporteurs of the working groups (Max Burton-Chellew, Thomas Valone, Paul Ewald, and Andrew King) deserve special recognition, for to draft a report during the Forum and finalize it in the months thereafter is never a simple matter. Finally, I extend my sincere appreciation to Luc-Alain Giraldeau, Philipp Heeb, and Michael Kosfeld: as volume editors, their commitment to interdisciplinary discourse ensured vibrant discussion and a successful Forum.

A communication process of this nature relies on institutional stability and an environment that encourages free thought. The generous support of the Ernst Strüngmann Foundation, established by Dr. Andreas and Dr. Thomas Strüngmann in honor of their father, enables the Ernst Strüngmann Forum to pursue its work in the service of science. In addition, the following valuable partnerships are gratefully acknowledged: the Scientific Advisory Board, which ensures the scientific independence of the Forum; the German Research Foundation, for its supplemental financial support; and the Frankfurt Institute for Advanced Studies, which shares its intellectual setting with the Forum.

Long-held views are never easy to put aside. Yet when this is achieved, when the edges of the unknown begin to appear and the resulting gaps in knowledge are able to be identified, the act of formulating strategies to fill such gaps becomes a most invigorating activity. On behalf of everyone involved, I hope that this volume will convey a sense of this lively exercise and promote further enquiry and cross-field interactions into the behavior of investors and exploiters.

Julia Lupp, Director Ernst Strüngmann Forum
Frankfurt Institute for Advanced Studies (FIAS)
Ruth-Moufang-Str. 1, 60438 Frankfurt am Main, Germany
https://esforum.de/

List of Contributors

Arbilly, Michal   Department of Biology, Emory University, Atlanta, GA 30322, U.S.A.

Barta, Zoltán   Department of Evolutionary Zoology, University of Debrecen, 4032, Debrecen, Hungary

Börner, Jan   Center for Development Research (ZEF), University of Bonn, 53113 Bonn, Germany

Brown, Sam P.   School of Biology, Georgia Institute of Technology, Atlanta, Georgia 30332–0230, U.S.A.

Burton-Chellew, Maxwell N.   Calleva Research Centre for Evolution and Human Sciences, Magdalen College, University of Oxford, Oxford OX1 4AU, U.K.; and Department of Zoology, University of Oxford, Oxford OX1 3PS, U.K.

Cardenas, Juan-Camilo   Facultad de Economía Universidad de los Andes Carrera, Bogotá, Colombia

Dall, Sasha R. X.   Centre for Ecology and Conservation, Cornwall Campus; and Biosciences College of Life and Environmental Sciences, University of Exeter, Penryn, Cornwall TR10 9EZ, U.K.

dos Santos, Miguel   Department of Zoology, University of Oxford, Oxford OX1 3PS, U.K.

Dubois, Frédérique   Department of Biological Sciences, Université de Montréal, Pavillon Marie Victorin, Montréal QC H3C 3J7, Canada

Ewald, Paul W.   Department of Biology, University of Louisville, Louisville, Kentucky 40292, U.S.A.

Foster, Gigi   School of Economics, University of New South Wales, UNSW Sydney, NSW 2052, Australia

Frijters, Paul   London School of Economics, Centre for Economic Policy, London WC2A 2AE, U.K.

Giraldeau, Luc-Alain   Faculty of Science, University of Québec at Montréal, Montréal, Canada

Greiner, Ben   Department of Strategy and Innovation, Wirtschaftsuniversität Wien (Vienna University of Economics and Business), 1020 Vienna, Austria

Hajjar, Reem   International Forestry Resources and Institutions (IFRI) Network, School of Natural Resources and Environment, The University of Michigan, MI 48109, U.S.A.; currently at the Department of Forest Ecosystems and Society, Oregon State University, OR 97331, U.S.A.

Heeb, Philipp   Laboratoire Evolution et Diversité Biologique (UMR 5174), Université Paul Sabatier, 31062 Toulouse, France

Herrmann, Markus   Department of Economics and CREATE, Université Laval, Pavillon J. A. De Sève, Québec, QC, G1V 0A6, Canada

Kacelnik, Alex   Department of Zoology, University of Oxford, Oxford, OX1 3PS, U.K.

Kameda, Tatsuya   Social Psychology, The University of Tokyo, Tokyo 113-0033, Japan

Khalmetski, Kiryl   Department of Economics and Center for Social and Economic Behavior (C-SEB), University of Cologne, 50923 Cologne, Germany

King, Andrew J.   Department of Biosciences, Swansea University, Swansea, SA2 8PP, Wales, U.K.

Kokko, Hanna   Department of Evolutionary Biology and Environmental Studies, University of Zurich, 8057 Zurich, Switzerland

Kosfeld, Michael   Economics and Business Administration, Goethe University Frankfurt, 60323 Frankfurt am Main, Germany

Leininger, Wolfgang   Economics and Social Sciences, TU Dortmund, 44227 Dortmund, Germany

Lotem, Arnon   Department of Zoology, Faculty of Life Sciences, Tel-Aviv University, Tel Aviv 69978, Israel

Mathot, Kimberley J.   Department of Coastal Systems Sciences, Royal Netherlands Institute for Sea Research (NIOZ), 1797 SZ 't Horntje, The Netherlands

McNamara, John M.   School of Mathematics, University of Bristol, Bristol BS8 1TW, U.K.

Mengel, Friederike   Department of Economics, University of Essex, Colchester CO4 3SQ, U.K.

Oldekop, Johan A.   International Forestry Resources and Institutions (IFRI) Network, School of Natural Resources and Environment, The University of Michigan, MI 48109, U.S.A.; School of Biology, Newcastle University, Newcastle NE1 7RU, U.K.; and Department of Geography, University of Sheffield, Sheffield S10 2TN, U.K.

Pauly, Daniel   UBC Fisheries Centre, The University of British Columbia, Vancouver, BC V6T 1Z4, Canada

Roche, Benjamin   IRD/UPMC UMMISCO, Research Institute for Development (IRD), 34394 Montpellier, France

Rustagi, Devesh   Management and Microeconomics, Goethe University Frankfurt, 60323 Frankfurt am Main, Germany

Sutherland, William J.   Department of Zoology, University of Cambridge, Cambridge CB2 3EJ, U.K.

Thomas, Frédéric   MIVEGEC, Research Institute for Development (IRD), 34090 Montpellier, France

Valone, Thomas J. Department of Biology, Saint Louis University, St. Louis, MO 63103, U.S.A.

van der Weele, Joël Department of Economics, Center for Research in Experimental Economics and Political Decision Making (CREED), 1001NJ Amsterdam, The Netherlands

Vollan, Björn Marburg Centre for Institutional Economics, Marburg University, Germany

Wedekind, Claus Department of Ecology and Evolution, University of Lausanne, 1015 Lausanne, Switzerland

Winterhalder, Bruce Department of Anthropology and Graduate Group in Ecology, University of California at Davis, Davis, CA 95616, U.S.A.

1

Introduction

Luc-Alain Giraldeau and Michael Kosfeld

Why do some individuals invest effort and time in making a resource available, while others simply wait and exploit the fruits of this effort? Why do some people cooperate in joint production and public goods provision while others free ride and abstain from contributing to aggregate welfare?

This phenomenon—the exploitation of others' investments, the free riding in the presence of others' cooperation—is studied by many scientists, including biologists, anthropologists, public health scientists, and economists. For one of us (Luc-Alain Giraldeau), as a starting PhD student in biology, it became the main theme of his research career some 36 years ago, when he stumbled onto a case of investor–exploiter interactions in pigeons.

At that time, Luc-Alain was trying to document how pigeons learned simple food discovery skills while foraging in groups, but after months of trials, this paragon of Skinner's operant conditioning seemed largely unable to learn even the humblest of tasks, like flipping lids or pecking on sticks to find food. To figure out why, Luc-Alain invested hours of painstaking observation, staring again and again at black and white videotapes. What he discovered was that the few pigeons that did learn allowed non-learning exploiters to feed off of their discoveries. Exploitation, it seemed, was the non-learners stumbling block to learning.

It did not take Luc-Alain long to realize that this form of exploitation had innumerable other consequences: on behavioral diversity within groups, on the efficiency of the group at discovering resources, aggression, food defense, spatial preferences, interspecific relationships, sociality, and social learning. As a biologist, he understood that exploitation was at the root of a wide range of phenomena and that it was remarkably ubiquitous. As the saying goes: "Once you see it, you just can't unsee it." Thus the rest of his research career was devoted to exploring the factors that governed the prevalence of exploitation within groups, mostly of birds. It quickly became obvious to him that animal exploiters might share commonalities with human exploitation or free riding in the economic world. He saw endless analogies between animal and human exploitation: Generic pharmaceutical companies exploit the discoveries of other pharmaceutical companies. Parents take advantage of other parents' investment in vaccinating their children and may decide not to vaccinate. Commercial fishing vessels use the success of others to decide where to fish.

As a biologist, he wondered whether these incidences simply bore a superficial resemblance to each other, or whether there was something more fundamental to link these human and animal systems? Could knowledge of animal systems inform researchers concerned with human affairs? Might discoveries in biology be useful to economists and public health policies? Might biologists improve their models of exploitation by incorporating advances developed by economists?

After spending a lifetime of research in the world of animal behavior, Luc-Alain was eager to learn whether this work could be translated to other fields, such as economics and public health. Equally, he was curious to see whether such translation could be multidirectional. So he approached the Ernst Strüngmann Forum to ask for their help.

Joining him in proposing this theme was another biologist, Philipp Heeb, who was instrumental in working on the initial proposal. Both served as cochairpersons of the Forum as well as members of the Program Advisory Committee, together with Alex Kacelnik, Michael Kosfeld, Julia Lupp, and Frédéric Thomas. Yet what brought Michael Kosfeld, an economist, to this discussion?

When Michael was initially approached to interact on this theme, the overall topic and objectives of the Forum immediately caught his attention. He well remembered how once, as a PhD student in economics with a prior training in mathematics, he had stumbled across one of the most famous games in economics: the so-called prisoner's dilemma. Originally discovered by Al Tucker to illustrate the potential nonsocial desirability of Nash equilibria, the game immediately fascinated him. On one hand, it was "theoretically obvious" that noncooperation is the only individually optimal strategy in this game, yet on the other, it seemed equally "empirically obvious" that cooperation is not simply a failure of human decision making but rather a very intuitive and natural behavior that is observed across many different contexts and situations. Since then, this game—in one form or the other—has influenced Michael's research, not only as a game theorist but also, in particular, as an experimental and behavioral economist.

Michael remembered the invigorating discussions that took place at the former Dahlem Konferenzen, during its 2002 meeting on "The Genetic and Cultural Evolution of Cooperation," chaired by Peter Hammerstein. There Michael engaged, as a postdoc, in a fascinating discussion of the "puzzle of cooperation" in both the human and animal world, together with a group of anthropologists, biologists, and economists. Because the Ernst Strüngmann Forum had resurrected the work of the Dahlem Konferenzen, Michael was truly excited to be involved in setting up a new interdisciplinary discourse that would scrutinize cooperation, this time from the perspective of exploitation.

Together, the members of the Program Advisory Committee created a framework for evolutionary ecologists, economists, anthropologists, and public health scientists to examine collectively whether there is some generality in the phenomenon of investment and exploitation as it manifests itself across a broad range of species and systems. Might understanding of this phenomenon, studied for decades along separate research traditions, be deepened through this interdisciplinary exchange? Could we come away from this discourse with new insights to tackle problems related to renewable resource management, public health, and institutional design?

The answer to both, of course, is a resounding yes...and no.

Challenges Posed by Interdisciplinary Traditions

An Ernst Strüngmann Forum is no stroll through the park. Intelligent, hardworking scientists with years of accumulated knowledge, reputation, and authority in a discipline are suddenly thrust together in the same space to discuss the very subject of their expertise with others who clearly haven't the faintest idea of who they are. Not an easy starting point for a group of career scientists!

As with any interdisciplinary exchange, initial interactions often aim at explaining, as simply as possible, one's own approach while listeners smile politely, often thinking: "This is pure gibberish." In French, such interactions are called *un dialogue de sourds* [a dialogue between the deaf]. The analogy is appropriate. Just as some noisy occupations make a number of their workers tone deaf, so too can scientists (as they become experts in their own discipline) become discipline deaf. This deafness stands as the main obstacle to any interdisciplinary exchange, and the Ernst Strüngmann Forum is no exception.

Speaking louder or yelling is of no help whatsoever at a Forum. One has only a short week to overcome one's hearing impairment, and this is no small task. However, given the Forum's experienced leadership and guidance, we are relieved to report that many were able to get past their sensory deficits in the course of this successful Forum.

Disciplinary deafness is not an act of bad faith. It is the result of accumulated hidden assumptions about the world which make up a field, determine its most important target questions, thus allowing everyone within it to understand each other. It can, in fact, even be a productive driver of within-discipline scientific excellence. As you read through these chapters, you will no doubt detect tension between biologists and economists. Zones of overlap and divergence between economists and evolutionary biologists appear much more clearly than they did back in November 2015, when we first met in Frankfurt.

In what follows, we highlight some of the issues that seem to be at the heart of this tension. Perhaps the best way to view them is as conceptual discontinuities between the fields which need to be exposed, so that areas of disciplinary deafness can be turned into a productive interdisciplinary dialogue.

Populations versus Individuals

Economists and evolutionary biologists both study exploitation and invest-ment, but they clearly have different criteria for what constitutes a satisfying answer. Put simply, evolutionary biologists find solace in population-level an-swers whereas economists seek individual-level explanations. For example, for an evolutionary biologist, there is no point asking why some individuals choose to invest: investors simply must exist, otherwise there would be no species. If no agent within a group ever invested in searching for food, then no one would ever eat and the whole group would die, and there would be nothing left to study. Investors, evolutionarily speaking, just have to be: end of story.

Economists, on the other hand, have a hard time accepting this population-level argument. They admit that if all members of a population free ride, then the collective outcome is worse, but in most economic examples the popula-tion will not die off or become extinct. Economists approach the issue from an individual level, and not just any kind of individual: one that is smart and has the capacity to plan ahead. For an economist, the real puzzle is: Why would anyone choose to invest in any costly effort, given that others will benefit as a result? In other words, economists develop models and experiments to address a question that evolutionary ecologists consider already answered.

Intriguingly, John Nash laid out this "duality" of a population- versus indi-vidual-level perspective in his original PhD thesis when he introduced the con-cept of (Nash) equilibrium in noncooperative games, a solution concept that has since influenced research in economics and other social sciences, probably like no other. Unfortunately, the population-level perspective, or mass action interpretation as Nash called it, was mostly forgotten during the rise of game theory in economics, while it became truly fundamental to biologists and is re-flected in concepts such as evolutionary stability, developed by John Maynard Smith and George Price. Only in the early 1990s did evolutionary game theory begin to partly enter economic modeling, but it seems true to say that it has never become part of mainstream economic thinking.

Producer–Scrounger versus Cooperation

Both evolutionary ecologists and economists agree that a population composed of all investors and no exploiters would typically be the best general outcome possible. Because evolutionary biologists do not concern themselves with the investor strategy, all of their work focuses on the causes of exploitation (free riding). Their models reflect this bias, typified most notably by the "producer–scrounger" game, which searches for the factors that govern the frequency of exploitation within a population: the way resources are distributed, their over-all abundance, whether resources can be defended or not, and whether agents are equal or unequal in strength.

For economists, to ask why free riding exists is a no-brainer: it is economically beneficial. The crux of the issue, then, is to find the conditions under which rational economic agents would choose to use the more costly but socially profitable alternative: invest. The games that place the emphasis on investment, such as the prisoner's dilemma game or public goods games, are framed in a perspective of cooperation, not exploitation. Hence the target questions in both disciplines are different and as a consequence so too is the terminology, and the models are framed in different traditions: exploitation in one, cooperation in the other.

Genes versus Learning and Culture

Evolutionary biologists focus on a wide range of species, many of which have no brains and little in terms of cognition. These species behave, nonetheless, and their behavior is far from irrational or erratic. Biologists, therefore, assume nothing about the thinking power of the agents they are studying. They simply take for granted that natural selection has endowed an agent with the necessary behavioral decision mechanism to make the best possible choice under a given set of circumstances. As a result, evolutionary ecologists employ game theory to study the interaction between organisms such as plants and their pollinators, without being the least bit concerned about the details of their decision mechanism. To evolutionary ecologists, the human brain, despite its complexity and immense capabilities in terms of culture and cultural transmission, must also be seen as an adaptive choosing device, nothing more. Accordingly, they assume that they do not need to concern themselves with the details of individual decisions, because a population will behave as expected by natural selection, on average.

Economists view the brain and humans quite differently from evolutionary biologists. As social scientists who focus exclusively on humans, the details of a decision are of paramount importance, and thus social scientists invoke (bounded) rationality, culture, social norms, legal and political institutions, enforcement rules, and reputation in their quest for answers. All of these, and much more, factor into their models and play a prominent role in the formulation of hypotheses. In addition, economists make a clear distinction between preferences and behavior. It may be rational for an individual to cooperate if s/he expects to be punished otherwise (say, by the police or by the courts), even though in general the individual prefers unilateral defection to joint cooperation. In contrast, an individual may rationally decide not to cooperate, even though s/he prefers joint cooperation to unilateral defection, if s/he believes that other players will defect as well (and there is no punishment). Exactly for these reasons, institutions—legal, social, or economic—are so important for economists, because they fix incentives, affect beliefs, and thus determine the explicit and implicit "rules of the game." The idea that evolutionary ecologists need not concern themselves with these factors can easily appear foolish and unrealistic to any economist.

Taxonomic Generality versus Specificity

Evolutionary biologists face a daunting task: they want to explain the behavior of a diverse set of taxa that adopt an unmanageable number of different behaviors. As a result, they tend to lump everything into a general, perhaps more superficial model that ignores specific cases but is useful in all circumstances. To do this, naturally, they must ignore a lot of what economists would consider to be important variables. The producer–scrounger game model provides a great example: it requires very few parameters, makes limited assumptions about the agents as well as the type of resource and so on. The model is meant to apply whenever a group of agents searches for a resource which, once found, can be shared. The game can be applied to bacteria-producing siderophores as well as whales hunting for patches of krill. Ecologists take great pleasure in knowing that their model can be applied widely to a broad set of species and problems.

The task faced by economists is equally daunting. They wish to account for decisions within a wide range of economic, social, and cultural settings in a single, but rather complex species: humans. Thus, they might not find it advantageous to change their models just to account for the behavior of another species, as their goal is to explain the behavior of humans in a given set of circumstances. To ecologists, economists thereby appear sometimes like compulsive hairsplitters: instead of coming up with one or two general models, they prefer to dig into institutional details, devising different games to capture more realistically all of the characteristics of the situation under study. Hence, while the behavioral ecologist keeps coming back with the general producer–scrounger game, the economist replies with a plethora of games, each for a slightly different way in which an exploitation situation can arise: the prisoner's dilemma, the linear or nonlinear public goods game, the common-pool resource game, and so on. Furthermore, economists get excited about the ways in which a given game can be modified even further (such that, e.g., individual incentives become aligned with social efficiency), for example, by writing legally binding contracts or by changing more generally underlying economic and legal institutions. This allows them to incorporate human-specific, and perhaps even cultural-specific, complexities to improve their predictions, but they do so at the price of taxonomic generalization.

The Structure of the Book

Whether you are an economist or an evolutionary ecologist, you will certainly appreciate the tension that loomed behind our discussions. However, you also need to understand the philosophy that underpinned our debate. Put simply, at an Ernst Strüngmann Forum hidden agendas are not tolerated, consensus is never a goal, and questions are viewed to be as important as answers. Within the confines of this intellectual retreat, participants are always encouraged to expose their individual and collective disciplinary ignorance, otherwise referred to as "gaps in knowledge." Once these become visible, the next challenge is then to brainstorm with colleagues on how such gaps could be filled.

This book is a summary of our debate and is structured around four main themes:

- Ecological and economic conditions of parasitic strategies
- Governance of natural resources
- Impact on human health
- Consequences for individual behavior, social structure, and design of institutions

Each section contains background information on fundamental concepts, which the Program Advisory Committee believed necessary for a fecund interdisciplinary discussion. Written before the Forum, these chapters were subsequently revised based on formal reviews as well as input received from the participants. As such, these chapters provide a glimpse of what economists, health scientists, or biologists initially considered important and may expose the disciplinary divides that exist. The final chapter in each section is a summary of each group's discourse. Drafted during the Forum by a heroic individual (the rapporteur) and written in collaboration with all group members, these chapters strive to overcome disciplinary deafness. Here you will find an economist trying to use the exploiter–investor terminology and a behavioral ecologist attempting to use economical terminology, e.g., in the characterization of resources in terms of rivalry and exclusivity. Genuine effort was made to bridge the gulf that separates disciplinary pursuit on the phenomenon of exploitation and here we wish to highlight the editorial efforts of Philipp Heeb and Julia Lupp who greatly helped us in trying to achieve this goal.

In this book you will find novel avenues for future research to address problems of sustainable use of renewable resources, the design of institutions, public health policies, and even medicine. Importantly, this volume represents the beginning of a further conversation that needs to be pursued between evolutionary biologists and economists, if we are to gain a better grasp of the factors that either govern the persistence of exploitation within social groups or which limit the use of cooperative investor strategies.

We are forever in debt to the Ernst Strüngmann Foundation and its Scientific Advisory Board for having afforded us this opportunity and providing the starting point for future interdisciplinary exchange. Let the dialog begin.

Ecological and Economic Conditions of Exploitation

2

Explaining Variation in Cooperative Behavior

Perspectives from the Economics Literature

Friederike Mengel and Joël van der Weele

Abstract

Casual observation and controlled experiments show that humans display great heterogeneity in their tendency to exploit others or invest in mutual cooperation. This chapter reviews models in the economics literature that can explain the coexistence of free riders (exploiters) and cooperators (investors). A distinction is made between models of full and bounded rationality. Although some models provide tentative explanations, there is a large gap between the empirical and theoretical literature, and there has been little effort to integrate long- and short-run models.

Introduction

Human societies thrive when individuals invest in cooperative relationships and common interest. At the same time, however, this situation offers opportunities for individuals to exploit the investments of others. Economists have studied the choice between investment and exploitation in the context of so-called social dilemma games, which juxtapose two kinds of actions. Individuals can "cooperate" with others by investing individual resources in actions that benefit the group. Alternatively, they can exploit, "defect," or "free ride" by choosing a strategy that benefits the individual but implies a material cost to the group.[1] This dichotomy offers a fruitful way to model many social and economic interactions that humans engage in on a regular basis.

---

[1] For the remainder of this chapter, we will use the terms "cooperation" and "defection" strategies in the dilemma situation that is the focus of the economics literature, and that fall within the broader concepts of "investment" and "exploitation." We will use the words "free ride," "defect," and "cheat" interchangeably.

Casual observation and experimental evidence show that there is great heterogeneity in people's tendency to cooperate or defect in social dilemma games (Ledyard 1995; Fischbacher et al. 2001; Camerer 2003). Explaining the origins and the stability of such heterogeneity is therefore an important task for social scientists. In this overview, we discuss whether economic models can help explain the origins and stability of heterogeneity. To do so, we distinguish between heterogeneity on different levels:

1. Long-term processes of natural selection, cultural transmission, and learning generate diversity in people's capabilities, knowledge, and preferences.
2. Institutional environments determine how such heterogeneity is expressed in behavior, depending on the incentives that such institutions provide.

Methodologically, both types of models are examples of game theoretic analysis, but the underlying assumptions are quite different. Long-run models rely on the assumption that strategies spread through social learning, cultural transmission, or the creation of offspring. These processes are often boundedly rational in nature. By contrast, short-run models typically assume rational or optimizing responses to the incentives provided by different institutions. This distinction is reflected in our discussion.

We begin with a discussion of how, for a given distribution of preferences, institutions that punish defection can induce heterogeneity in behavior. Thereafter we turn to models of bounded rationality and discuss how learning, imitation, and evolution may contribute to equilibria with heterogeneity in preferences or strategies. Throughout, we focus on stylized social dilemma situations, whereas Oldekop and Hajjar (this volume) focus on the importance of contextual factors.

The distinction between long-run processes of preference formation and short-run reactions to incentives is not always clear-cut. Institutions that incentivize certain behaviors in the short term can also cause long-term changes in preferences and behavior (Bowles 1998). Conversely, long-term movements in preferences will change the kind of institutions that will be necessary to promote cooperation. Where available, we discuss a literature that addresses this two-way relationship and argue for the need for more comprehensive theories.

Punishment Institutions

Social dilemma games (e.g., prisoner's dilemma, public goods game) exemplify the trade-off between exploitation and investment that is the focus of this volume. In such dilemmas, individuals choose between a personally costly "cooperative" action that benefits other group members and "free riding" or "defection," which maximizes the material payoffs to the individual at the expense of the group.

All individuals in a group are better off when they can achieve cooperation from all members, compared to a situation where everyone defects. To limit defection, societies have developed a diverse set of institutions to punish defectors. However, punishment may not deter all defections, thus allowing both cooperative and free-riding behavior to coexist. Here we review evidence for the coexistence of both behaviors in institutional contexts that have been analyzed in the economic literature. While we focus on the effect that institutions have on the behavior of optimizing agents, we also highlight how institutions could affect long-term processes of preference formation.

Centralized Punishment

Punishment by a central authority has been a major topic in the philosophical literature since at least Thomas Hobbes. In the seventeenth century, Hobbes argued how a government or Leviathan could improve the situation of noncooperating individuals in a "state of nature" and be sustained as part of a social contract between individuals. An economic take on this idea is the economic model of crime, formalized by Becker (1968). The starting point in this model is a society of agents who have the opportunity to take an action (crime) that brings personal benefits but hurts others in the society. Typically individuals will differ in the personal cost and benefits of crime, due to differences in moral convictions, wealth, and other personal circumstances. Potential criminals rationally weigh the costs and benefits of the crime. A central authority can influence the calculations of these individuals and improve efficiency by raising the cost of crime, for example, through sanctions or imprisonment. Thus, the model of crime extends "the economist's usual analysis of choice" (Becker 1968:170), by analyzing crime as a good and punishments as that good's price.

Becker showed that it is not optimal to deter all crime when deterrence is costly and the costs and benefits of crime differ between individuals. Rather, crime should be deterred only up to the point where the marginal costs of enforcement plus the marginal benefits to the criminal equal the marginal cost of crime to the victim. A pragmatic, optimizing authority will allow some crime to occur, either because it is too costly to eradicate or because it is relatively harmless to the victim, or both. Thus, the optimal policy that flows from this model results in the coexistence of both compliant and criminal behavior.

There are many subtle forms of exploitation or free-riding behavior that are technically not "crimes" punishable by law. Becker's model is very general and can be applied to these subtle forms of defection. Similarly, there is flexibility in the incentives and the authorities that can be considered. For example, the management of a firm can discourage shirking behavior by paying part of a salary in bonuses for high performance, or a football coach can bench players who perform poorly.

The central implication of this theory is that crime should fall with both higher penalties and a higher probability of getting caught. There is a decade-old debate about the empirical validity of this "deterrence hypothesis." Although there is some evidence that the probability of getting caught matters, there is no consensus as to whether higher penalties reliably deter crime. The effects of deterrent policies appear to be highly dependent on the social context (van der Weele 2012a).

In the long term, centralized punishment can also affect the coexistence of cooperative and selfish "types" who differ in their preferences for cooperation. Consider, for example, interactions involving bilateral exchange, where one party can cheat the other party (Huck 1998). Trading partners cannot distinguish between cooperative and defector types, so the latter will take advantage to cheat on their contracts. In this situation, probabilistic detection of cheating associated with penalties favors cooperative types who are more likely to comply. As a consequence, cooperative types become more prevalent in the population, causing the optimal size of sanctions to decrease over the long term.

By contrast, when the type of the interaction party is observable at least with some probability, the imposition of sanctions can favor defectors. In the absence of sanctions, cooperative types would never interact with defectors, who would suffer low payoffs and make up only a small share of the population. Bohnet et al. (2001) showed that the presence of enforcement weakens this kind of ostracism and leads to an increase in the population share of defectors, as long as penalties are relatively low. Thus, cooperation or trustworthiness is crowded out with weak enforcement, but crowded in with strong enforcement. Indeed, in their experiment, Bohnet et al. (2001) found that weak penalties lead to an increased prevalence of cheating.

In summary, coexistence of cooperators and free riders exists naturally in a world with costly enforcement and heterogeneity in the costs and benefits of crime. The long-run effects of centralized enforcement depend on the size of the penalties and the available information about the interaction partner.

Decentralized Punishment and Social Norms

Many forms of punishments are not carried out by a central authority, but rather by a community of peers.[2] Peers have an important advantage over a central authority in that they will often have more information about the nature of transgressions and their perpetrators. Peer punishment can take the form of

---

[2] The conceptual distinction between centralized and peer-sanctioning schemes does not mean that the two are independent of each other. For example, Huck and Kosfeld (2007) argue that when an authority reduces sanctions to discourage crime, this may also lead to the abandonment of neighborhood watch groups engaged in peer surveillance. Van der Weele (2012b) shows that when a government has superior information about the number of free riders in society, it may refrain from setting severe penalties to avoid signaling to citizens that being a free rider is the social norm.

tacit or open disapproval, withholding of cooperation, ostracism, or even physical attack against the perpetrator. Peer punishment, however, is often costly to the punisher, and the enforcement of cooperative social norms becomes itself a public good, often referred to as a "second-order public good problem."

The second-order public good problem is easier to solve than the first-order problem, as a large part of the population seems willing to punish transgressions at a potential cost to themselves. Fehr and Gaechter (2000), as well as many follow-up studies, demonstrated this experimentally in the context of a public goods game, a multiperson version of the famous prisoner's dilemma. The authors augment this game with a punishment phase, in which participants can take away earnings of noncontributors at a cost to their own experimental earnings. Many participants do indeed choose to punish which, despite the initial destruction of resources, leads to higher cooperation rates and higher efficiency over time (Gaechter et al. 2008).

The cost and effectiveness of peer punishment is likely to depend on the ratio of defectors and cooperators. When cooperators are relatively numerous, punishment resources can be concentrated on a smaller sample of defectors, leading to higher (probability of) punishment. At the same time, an increased expectation of punishment is likely to lead to a higher cooperation rate.

We show the consequences of such punishment complementarities for the coexistence of free-riding and cooperative behavior in a canonical model. Variations of this model have been applied by different authors to different instances of defection such as tax evasion (Lindbeck et al. 1999; Traxler 2010). Suppose that there is a community of a countable infinite number of agents indexed $i = 1, 2, \ldots$. Each agent can choose to behave cooperatively or to defect. Payoffs from cooperation are zero; payoffs of defection are given by $\theta_i - C(n)$, where θ_i is the "type" of agent i that encapsulates all psychological and material payoffs from defection specific to that agent. Agents differ with respect to their θ_i and are distributed over the type space according to some distribution $F(\theta)$ with full support on $[0, \theta_{max}]$. The second term, $C(n)$, represents the social cost of punishment to the defector. This cost depends negatively on the fraction n of the population that defects, reflecting the assumption that the more defectors there are, the more punishment will be diluted.[3]

A rational individual thus defects if and only if $\theta_i \geq C(n)$. If we define by θ^* the type that is just indifferent between defecting or not, then all types $\theta > \theta^*$ will defect. Since $n = 1 - F(\theta^*)$ is the fraction of defectors, it follows that θ^* is defined implicitly by the equation $\theta^* = C(1 - F(\theta^*))$. Under suitable conditions on the functions $C(n)$ and $F(\theta)$, this model yields a situation with multiple equilibria, as illustrated in Figure 2.1.

---

[3] To keep things simple, we have assumed here that the payoffs of defection (and the effectiveness of punishment) do not depend on n. If they do, one can get similar qualitative results, as long as these payoffs of defection fall less quickly with n than its costs.

F. Mengel and J. van der Weele

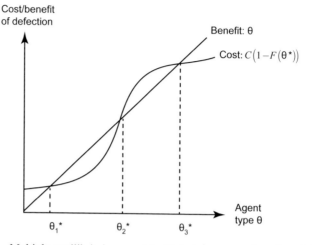

Figure 2.1   Multiple equilibria in a model with heterogeneous benefits of defection. The x-axis shows the agent type θ; the y-axis shows the costs $C(1 - F(\theta^*))$ and benefits (θ) from defection. Equilibrium is found where the two lines intersect, and costs and benefits are equal.

Equilibria characterized by high levels of defection and low levels of peer enforcement (e.g., θ_1^* in Figure 2.1) coexist with equilibria featuring low levels of defection and high levels of enforcement (e.g., θ_3^* in Figure 2.1). Depending on the shape of the functions $C(n)$ and $F(\theta)$, there may be a large number of such equilibria. Note that the equilibrium associated with θ_2^* in Figure 2.1 is unstable. That is, suppose that type θ_2^* deviates from this equilibrium and chooses to cooperate. This raises the cost of defections for the remaining agents, making it optimal for some types $\theta > \theta^*$ to cooperate as well, and causes the equilibrium to unravel. Conversely, if some types just below θ^* defect, it becomes optimal for "lower" types to do so as well. Thus, θ_2^* is better thought of as a "tipping point" rather than an equilibrium, on either side of which a different stable equilibrium becomes an attractor.

Common terminology associates the (equilibrium) level of n with a "social norm," because it measures the degree to which cooperation and defection are normal actions in the population. Glaeser et al. (1996) argued that variation in social norms can explain the lion's share of the empirical variation in crime. In their intercountry comparison, using data from 1980, the homicide rate in the United States was about 150 times higher than in Japan. Looking at differences between U.S. cities, Glaeser et al. found that the crime rate in Atlantic City, New Jersey, was about 400 times higher than the nearby city of Ridgewood Village. On an even more detailed intra-city level, the crime rate in the 1st Precinct of New York City was about 10 times greater than in the 123rd Precinct. Examining the crime statistics in New York in more detail, the authors show that at most 30% of these differences can be accounted for by observable differences between different locations (e.g., levels of income, schooling, female-headed households, arrest rates). The rest, they argue, is due to peer effects or social norms.

In summary, complementarities in the effectiveness of peer punishment imply the existence of multiple equilibria or social norms. One can think of these multiple equilibria as "parallel societies" that may exist in different places or at different times. When there is multiplicity, theory cannot predict *ex ante* how a given distribution of preferences translates into cooperative behavior, but each equilibrium may itself involve a stable coexistence of cooperation and free-riding behavior.

Exclusion and Sorting

A particular form of peer punishment is ostracism or exclusion. Forming and terminating relationships can be part of the strategy set, but separation can also arise as an equilibrium phenomenon even if sorting operates via other mechanisms. Our focus in this section is on the latter type of models; an example of the former is taken up later.

Kosfeld et al. (2009) derived a good example of a mechanism that relies on sorting through their analysis and experimentally testing a coalition formation model within the context of public good provision. They modeled institution formation as a three-stage game. In the first stage, each player decides whether to participate in an organization that, once implemented, exerts punishment on individual members who do not contribute their full endowment to the public good. The organization is costly, and only players who are members of the organization can be punished. In the second stage, players learn how many of the other players are willing to participate. The organization is implemented if and only if all players willing to participate agree to its actual formation. In the final stage, the public goods game is played. Theoretically, two types of equilibria can be sustained: one in which at least a minimal amount of players establish an institution and contribute to the public good; the other where no institution is established. Under the first type of equilibria, coexistence can be established.

Models of coalition formation have also been used to explain cartel formation (Green and Porter 1984). The incentives in these cases are very similar, but most would view the successful establishment of a collusive institution as less beneficial, because of the harm inflicted on third parties, often consumers.

Given the intuitive appeal and wide range of applications of such models, it is not surprising that these ideas have been tested in a variety of experiments. Kosfeld et al. (2009) tested their mechanism in the lab and found that participants are unwilling to support institutions where some individuals can free ride. The vast majority of outcomes found in the lab feature institutions where everyone is part of the institution and contributing, or where no one is. Guererk et al. (2006) found evidence of the effectiveness of sorting in a slightly different setting. In their experiments, participants endogenously choose to select into a punishment mechanism. The main difference to Kosfeld et al. (2009) is that those who select into the punishment mechanism do not interact in the public goods game with those who did not select into the mechanism. While sorting into the punishment institutions can sustain cooperation in the lab in both studies, it is full cooperation, not coexistence, that is sustained.

In summary, some models have been able to establish coexistence via sorting or exclusion mechanisms. Notably, however, when tested in the lab, coexistence has not been established. Whether these lab results (typically obtained in small groups of 3–5 participants) extend to much larger groups remains to be seen.

Reputation Motives

In our discussion of punishment, we have implicitly assumed that defectors can be identified, at least with some probability. However, the information that is available about a person's past behavior depends itself on institutions. For example, one of the crucial obstacles facing the development of online commerce platforms, such as eBay, is the development of reputation formation mechanisms that allow customers to identify fraudulent sellers.

Researchers have modeled the effect of reputational motives on cooperation in various ways. Kreps et al. (1982), for example, investigated the power of reputational motives for cooperation in the finitely repeated prisoner's dilemma. Under common knowledge that all agents are selfish, theory predicts that cooperation will unravel: since cooperation is an irrational move in the last round of play, promises based on cooperation in later rounds have no credibility. Kreps et al. (1982), however, posit that players believe that, with some probability, their opponent may be a "tit-for-tat" player (cooperate on first move and then match the strategy opponent used on last play). If such beliefs are sufficiently high, they show that it may be rational to pretend to be a tit-for-tat type, at least until the end of the game draws near, so as to exploit the other player's conditional willingness to cooperate. As a consequence, all players may act cooperatively, even though none were of a tit-for-tat type.

Economic experiments that test this reasoning show that there is substantial heterogeneity in cooperative behavior. For example, Andreoni and Miller (1993) conducted a prisoner's dilemma game where the same partners interacted repeatedly for a finite number of rounds. They observed a cooperation rate of about 60% in early rounds, which deteriorated toward the end of the game. By contrast, when reputation formation was not possible, the cooperation rate was much lower at about 20%. In a similar experiment carried out by Bolton et al. (2004), sellers and buyers interacted repeatedly, and sellers had the possibility to defraud the buyers. Bolton et al. included a feedback condition: partners rotated but buyers could observe the behavior of the buyer in previous rounds. Although the conditions where participants played with the same partner in each round yielded almost full (about 90%) trustworthiness from sellers, trust in the feedback conditions hovered around 70% until the last few rounds of the experiment.

The model by Kreps et al. (1982) does not explain why some people behave like cooperative partners while others do not, but it provides a starting point. One plausible explanation is that participants have different initial beliefs about the likelihood of facing a cooperator, although it is unclear exactly where these beliefs originate, as the experimental conditions were the same for all. Since the theory allows for both defection and cooperation as an equilibrium, it may also be that different experimental subjects play different equilibria, without managing to converge on a single equilibrium. Finally, some forms of learning may rationalize these results (see next section).

In the long run, reputation motives can lead to the evolution of strategies of "indirect reciprocity." In the image scoring game introduced by Nowak and Sigmund (1998), two players are randomly drawn in each round from a population: one player is randomly selected to be the donor and the other the receiver. The donor can decide to "keep," yielding a payoff of c to the donor and 0 to the receiver, or the donor can "give," yielding a payoff of $b > c$ to the receiver and 0 to the donor. The donor's decisions result in an "image score" that is visible to the partner: it is 1 if the donor gave at the last opportunity, and 0 otherwise.

Nowak and Sigmund (1998) considered the strategies employed by universal defectors (who never give), universal altruists (who always give), and "discriminators" (who only give to a partner with a sufficiently high image score). Such discriminators, in fact, practice a form of indirect reciprocity; that is, they cooperate in the hope that the resulting image will induce cooperation from a future interaction partner. Nowak and Sigmund demonstrated that in this model, discriminators can successfully invade a population of universal defectors when the likelihood of knowing the partner's image score q exceeds the ratio c/b. However, universal altruists can invade discriminators, who will occasionally resort to punishment at a cost to themselves. Altruists can be invaded, in turn, by defectors, causing a never-ending evolutionary cycle.[4]

In summary, in finitely repeated dilemma games with reputation formation, a mixture of cooperation and defection can be observed. Economic theory suggests that this is due to differences in beliefs about the behavior of other agents, but it does not explain the origins of such beliefs. Evolutionary models link reputation formation to indirect reciprocity, but these models are hard to test. Further work is thus needed to determine how the increased importance of reputations, in a world with social media and an increasing amount of available information, impacts cooperative behavior.

---

[4] Note, however, that Lotem et al. (1999) showed that if there is a steady exogenous supply of defectors, the discriminator's advantage over altruists remains, and the frequency of discriminators stabilizes.

Bounded Rationality

In this section, we focus on explanations under which boundedly rational agents learn to adopt certain behaviors or social norms via processes of social learning or cultural transmission. Boundedly rational agents use heuristics or learning rules which may ultimately make them fail to appreciate possible private gains. Thus, under this class of explanations, punishment is not required to sustain cooperation and/or exploitation.

Learning

Models of learning specify intuitive updating and choice rules under which agents do not always rationally incorporate all available information. Learning rules differ widely, according to their degree of sophistication, and range from very simple reinforcement type rules to forward-looking and optimizing agents. Here we do not attempt to give a complete overview of the learning literature (for a good overview see Dhami 2016, Chapter V). Instead we focus on select models that can explain the coexistence of cooperators and free riders.

Reinforcement, Endogenous Aspirations, Forward-Looking Learners

In behavioral psychology, reinforcement increases the frequency of a certain behavior whenever that behavior is followed by a stimulus that is appetitive or rewarding. In economics, reinforcement learning refers to the fact that an agent's propensity to choose any given action will be proportional to past payoffs obtained with that action (Roth and Erev 1995). Payoffs are thus the economic agents' antecedent stimulus and motivator in these models. Standard (Erev–Roth type) reinforcement learning models are approximated by the evolutionary replicator dynamics and can only support Nash equilibria (and hence defection in the prisoner's dilemma) as stable states (Börgers and Sarin 1997). Since Nash equilibrium is a coarsening of evolutionary stability (Maynard-Smith and Price 1973), this also means that cooperative outcomes are not evolutionarily stable.

Some variants of reinforcement learning with endogenous aspirations—where agents are satisfied when their aspiration is met and aspirations adjust to recent payoff experiences (Simon 1956)—have been shown to support cooperation (Karandikar et al. 1998). In the following scenario, studied by Karandikar et al. (1998), players are randomly matched to play a 2 × 2 prisoner's dilemma game. Each player has an aspiration at each date and takes an action. The action is switched at the subsequent period only if the achieved payoff falls below the aspiration level, with a probability that depends on the shortfall. Aspirations are updated in each period, depending on the divergence of achieved payoffs from aspirations in the previous period. Karandikar et al. (1998) showed that if the speed of updating aspiration levels is sufficiently slow, then the outcome, in the long run, must involve both players cooperating most of the time. While there is no coexistence of cooperators and defectors in a stable state, players may (and occasionally do) profit by deviating from cooperative behavior. The dynamics of the process, however, ultimately leads back to mutual cooperation.

In another class of models, where it is assumed that agents are forward looking (anticipating future path of play) but still adaptive (learning from past experience), learners are much more sophisticated. With forward-looking agents, cooperation can be sustained in finitely repeated interactions because agents learn that histories involving defection are more often followed by defection than histories involving cooperation. Forward-looking agents learn that defection can sour relationships. Such forms of learning can explain why there is more cooperation when reputation formation is possible (Andreoni and Miller 1993; see also above discussion on reputation motives).

Imitation, Networks, and Exclusion

Let us now turn to social learning models, where agents copy behaviors observed in others, and revisit exclusion as a mechanism to sustain cooperation. This time, though, sorting is not a choice by optimizing agents; it arises endogenously via a process of imitation learning by boundedly rational agents.

Similar in spirit to the group/kin selection literature in biology, a number of models have been proposed that rely, to some extent, on excluding defectors from beneficial interactions with cooperators. These models exploit the idea that if, for whichever reason, cooperators were to interact more frequently with other cooperators (and defectors with defectors), then cooperators could achieve higher evolutionary fitness, precisely because joint cooperation pareto-dominates joint defection; that is, all players are better off under joint cooperation compared to joint defection. This idea has appeal because it is grounded in evolutionary considerations and does not rely on the introduction of "types" (or tweaks to payoffs) that would not survive the test of evolutionary fitness.

Imitation learning, where agents imitate successful behavior of others, has been able to produce stable states with coexistence of cooperators and defectors. In a seminal paper, Eshel et al. (1998) modeled agents located on a circle network imitating the successful actions of their neighbors. Agents in this model copied an action if and only if it yielded higher average payoffs in their neighborhood. This is very different from blind copying, customarily used in the behavioral ecology literature (see also Dubois et al., this volume). Under the assumptions made in Eshel et al. (1998) there can be clusters of cooperators coexisting with clusters of defectors in the network. Cooperators in the middle of a cluster or defectors in the middle of a cluster of defectors will not change their behavior via imitation learning simply because everyone they observe (and hence could imitate) chooses the same action as themselves. What about the cooperators and defectors at the fringes of these clusters? The defectors are getting a better payoff than the cooperator they observe (who they exploit) and thus will not be tempted into imitating. The cooperators also will not switch because, in these equilibria, they observe other cooperators (those in the center of a cluster) who are very well off. These equilibria, however, have proven to be very fragile. In subsequent literature it has been shown that they are not obtained in other networks and that allowing agents to use information from agents further away (even if only second-order neighbors) destroys the result even in the circle (Goyal 2007; Mengel 2009).

In a model of endogenous network formation, the possibility of coexistence of free riders and cooperators has been demonstrated by Fosco and Mengel (2011). With endogenous networks, people imitate actions *and* link choices of successful others; this leads to a coevolution of the network with choices in the prisoner's dilemma. In absorbing states of this coevolutionary process, the shortest path between any two cooperators never involves a defector. The reason is that any two cooperators separated by a defector will want to sever ties with the defector and establish instead mutual links. This leads networks to form dynamically, as illustrated in Figure 2.2: cooperators end up occupying central positions in the network, with free riders in more peripheral positions (Figure 2d).

Free riders will not imitate cooperation, since the only cooperators they observe are linked to many defectors and make poor payoffs. Most cooperators do not observe any defectors. Those that do observe defectors are by and large linked to other defectors, thus making lower payoffs than the cooperators

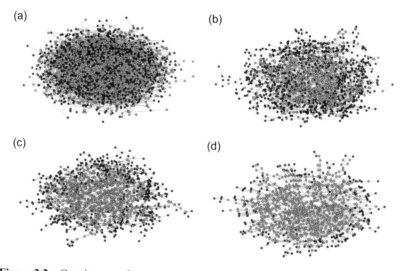

Figure 2.2 Coexistence of cooperators (green nodes) and free riders (red nodes) in a model of endogenous network formation. Initially (a) cooperators and free riders are randomly allocated on the network. (b)–(d) As agents start to form and sever links, free riders are pushed increasingly toward the periphery of the social network.

they observe in the periphery. Here a key element of bounded rationality is that these "bridging cooperators" (who are linked to both defectors and cooperators) assess the benefits of cooperation by comparing themselves to people in very dissimilar situations.

There is no clear empirical evidence on whether defectors or cooperators tend to be more central in social networks, though some studies have found that altruists (measured by giving in the dictator game) tend to be more central (Branas-Garza et al. 2010). In a study of smoking behavior of 12,067 people assessed between 1971 and 2003, Christakis and Fowler (2008) found that smokers moved increasingly to the periphery of their social networks. They interpret this to reflect a societal change in perception, where over time, smoking came to be seen as antisocial. These models may also provide some insight into the interactions of hunter-gatherer groups, which were mostly bilateral in kinship, as opposed to the lineage- or village-based grouping of horticulturalists and agriculturalists. Conversely, allowing for kinship structure in some of these models might generate new insights.

In summary, while various classes of individual learning models can only sustain Nash equilibria (and thus defection), models using endogenous aspirations as well as models that rely on limited forward-looking players can sustain some degree of cooperation. None of these, however, has been shown to sustain stable coexistence. By contrast, some social learning models, especially those which rely on imitation, have produced "proper coexistence" of cooperators and defectors who interact with each other in a social network.

Categorization

An important aspect, which economic theories of learning have only started to incorporate recently, is that agents' experiences and learning are not grounded in a single game alone: they develop across a great variety of situations. Theories of categorization and learning across games model how agents' learning and behavior in one situation will be affected by others. Categorization occurs to economize on reasoning cost or to make faster decisions. Allport (1954) famously noted that "the human mind must think with the aid of categories. We cannot possibly avoid this." In the prisoner's dilemma, agents may cooperate because to do so seems optimal *on average* across a broad category of different situations.

There is only a limited literature available on categorization and cooperation in economics. However, recent advances have been made in understanding when categorization may occur and in showcasing examples of categorization affecting behavior in situations that resemble social dilemmas.

Noting that humans compete in "Machiavellian tournaments," Samuelson (2001) modeled situations where agents put increasing amounts of effort into the tournament and lump together other decision situations in coarse categories as a consequence. Using this approach he showed that fair splits in ultimatum games can be sustained in equilibrium as players will lump ultimatum games together with (longer) alternating bargaining games, where fair splits are equilibrium outcomes. More generally, bunching can be advantageous in an evolutionary sense whenever reasoning or other costs make it prohibitive to devise strategies for each game separately. In such settings cooperation can be sustained in games where the incentives to defect are not "too big," because these games are bunched together in evolutionary equilibrium with others where cooperation is a Nash equilibrium.

Another strand of literature has focused on coarse beliefs. Jehiel (2005) developed a concept called analogy-based expectations equilibrium wherein people form the same beliefs across different situations bunched together in the same equivalence class of games. Irrespective of how beliefs are formed, there are no beliefs that can rationalize choosing a dominated strategy, such as cooperation in the prisoner's dilemma. Early experimental evidence supports this point, demonstrating how categorization can affect equilibrium play in a range of games, with the exception of those that have dominant strategies (Grimm and Mengel 2012).

This relatively young research area could be a promising avenue for further research, possibly studied in conjunction with identity concerns. People interact in a great variety of situations and how we partition and view the world is part of our culture and social identity. Existing studies have shown that taking these factors into account can rationalize seemingly irrational behavior in experimental games that are closely related to social dilemmas (Jehiel 2005; Mengel 2012). Exploring more deeply such identity-driven motives in conjunction with coarse (culturally determined) reasoning could lead to novel results on coexistence.

Cultural Transmission and Evolution of Preferences

Culture is transmitted within and across generations. The cultural transmission literature tries to model this process, often focusing on the subtle interactions between genetic and cultural evolution thought to occur at differential speeds. Models of cultural transmission distinguish between three types of transmission (Cavalli-Sforza and Feldman 1981; Boyd and Richerson 2005): (a) horizontal transmission, including the formation of social norms and peer pressure, (b) vertical transmission, including parent's socialization efforts, and (c) oblique transmission through, for example, media, teachers, or other role models.

One of the most interesting approaches in this area focuses on the endogeneity of preferences and stems from the work of Bisin et al. (2004; see also Bowles 1998). Building on earlier work by Cavalli-Sforza and Feldman (1981), Bisin et al. focused on vertical transmission, assuming that parents who socialize their children are motivated by altruism (i.e., a non-selfish concern for their children's well-being). However, altruism is not perfect in the sense that parents evaluate their children's payoffs with their own preferences (rather than their children's). Parents invest some effort into socializing their children: if they succeed, the child will adopt the parents' preferences; if they fail, horizontal transmission will kick in and the child will become socialized according to the preferences of the majority of the child's peers. Altruistic preferences survive as a minority preference, because minorities have higher incentives to socialize their offspring to their own preferences than majorities do. Thus, this is one of the few true coexistence results in the literature. Empirical evidence on how altruistic behavior is affected by the minority or majority status of a group remains scarce.

In models of horizontal transmission, conditionally cooperative preferences have been shown to coevolve with matching structure. Here, unlike in the models discussed earlier (see discussion on exclusion and sorting), the mechanism is not so much one of exclusion, where the fact that some agents are excluded from beneficial interactions with cooperators leads to coexistence. Instead, the strength of moral concerns itself changes with the matching structure (Mengel 2008). The idea is that people feel guilty about free riding, for example avoiding taxes, if these moral concerns are shared by many others in their social environment. If, by contrast, most people do not feel guilty about free riding, then feelings of guilt will be subdued (see also above discussion on exclusion and sorting). The conditionality of the moral concerns gives rise to conditionally cooperative behavior (Fischbacher et al. 2001). Since conditional cooperators adapt their behavior with the share of free riders in their environment, they cannot be easily exploited. This underlies the evolutionary fitness of cooperative behavior. While the previous mechanism works without observing other's preferences prior to interaction, a simpler mechanism that does require observability of preferences was noted by Bester and Gueth (1998). They note that if preferences can be observed, then conditional cooperation can be evolutionarily stable whenever there is enough "strategic complementarity" in decision making. Conditional cooperators cooperate whenever they observe others with cooperative preferences; otherwise, they free ride.

To summarize, mechanisms of cultural transmission have produced coexistence with free riders and cooperators interacting. The results rest on either of two mechanisms: vertical transmission with endogenous socialization efforts by parents, or horizontal transmission with endogenous norm strengths. Currently there are no convincing empirical tests of either mechanism, thus leaving a possible avenue for further research.

Discussion and Conclusion

We have surveyed the economics literature in search of explanations for the coexistence of free riders and cooperators in social dilemma situations. In doing so, we distinguished between theories which take the heterogeneity of agents as given and focus on how institutions guide the behavior of rational agents, and theories which try to explain the origin of preferences in processes of bounded rationality.

When it comes to punishment institutions, some naturally allow the expression of heterogeneous preferences into heterogeneous behavior. For example, when enforcement costs are positive, it is inefficient to deter all crime, leading to coexistence of cooperative and free-riding behavior. Complementarities in peer punishment lead to multiple equilibria, which can be interpreted as "parallel societies" governed by different social norms. Another set of theories rely on exclusion mechanisms, where cooperators manage to generate surplus among themselves, despite the existence of free riders.

In terms of models of bounded rationality, a few theories have been successful in explaining "proper coexistence," where defectors and cooperators interact in stable proportions. Again, some of these models rely on the ability to exclude defectors and marginalize them to the fringes of the network (Fosco and Mengel 2011). Another promising avenue is cultural transmission mechanisms, where minority groups sustain themselves through intense socialization efforts by parents (Bisin et al. 2004).

These models provide starting points to think about coexistence. However, there are several reasons why they do not represent a unified body of research. First, there is a substantial gap between theory and empirical work. Theories of learning or cultural transmission predict long-run dynamics that are hard to isolate from other factors. As far as we know, no empirical validation exists for such models. Some empirical research has demonstrated cultural transmission (Dohmen et al. 2012), but testing of theoretical work is limited. Theories of the effects of institutions are easier to test (e.g., in the laboratory), but such tests have so far not produced valid explanations of coexistence. For example, theories of reputation formation predict either universal cooperation or noncooperation, but not the intermediate outcomes observed in actual experiments. Conversely, models of coalition formation, such as by Kosfeld et al. (2009), predict coexistence but the experiments find mostly full cooperation.

Second, this literature has not accounted for evolutionary foundations. While investigating economic institutions, economists simply assume heterogeneity in preferences for the payoffs of others, feelings of warm glow, altruism, or reciprocity. These "social preference" models have remained ad hoc, and there has been little discussion of their evolutionary origins (although see Alger and Weibull 2013). Just as social preference models lack evolutionary foundations, models of the dynamics of learning and evolution have paid little attention to the effect of legal or cultural institutions, although interesting leads are emerging that evaluate the effects of institutions on preferences for cooperation (Bowles and Polania-Reyes 2012). Clearly, future research needs to examine the two-way interactions between institutions and the development of individual tastes. Such models could help to explain the emergence and effects of institutional variations, like religion and government, and to evaluate the long-term effects of economic and political shocks.

Finally, there is ample evidence that social and moral preferences are more complex than assumed in the most popular economic models. The same individuals often behave quite differently in different dilemma situations (e.g., Blanco et al. 2011), and there are important spillover effects between behavior in different environments (e.g., Grimm and Mengel 2012). Thus, it is not clear how stable traits are over time and across different contexts. Models based on the assumption that people intrinsically favor altruistic behavior are unable to explain evidence that behavior that is "too altruistic" is viewed negatively and sometimes even punished (Herrmann et al. 2008), the fact that the framing of a decision problem often matters (Andreoni 1995), or that people seem to care about expectations of others (Charness and Duwfenberg 2006). Some attempts have been made to represent the complex psychology of social and moral concerns in game theoretical models, often involving more parameters and more complex equilibrium concepts (Bénabou and Tirole 2006). Without evolutionary foundations or other ways to pin down parameters, such models may lack parsimony and predictive power.

In the process of writing this review, we found that explaining coexistence of free riders and cooperators has not been a goal of the economics literature. The literature summarized in this chapter focuses on explaining a particular behavior or trait, such as altruism or indirect reciprocity, or the impact of some institution, like climate coalition or a punishment scheme. Coexistence sometimes emerges as a byproduct of these endeavors, but it was never the direct object of inquiry.

To make progress, this will have to change. The economics literature could get inspiration by comparing models of social dilemmas in economics with behavioral ecology models of producers and scroungers. While both are models of investing and exploitation, coexistence is an equilibrium outcome in the latter, but not in the former, where the only Nash equilibrium (and hence also the only evolutionary stable state) involves free riding (see Burton-Chellew et al., this volume). More generally, we hope that the discussions brought forth in this volume will help both disciplines to exploit their common interest and help economists invest in new explanations for coexistence.

Acknowledgments

We thank Sam Brown, Luc-Alain Giraldeau, Philipp Heeb, Kiryl Khalmetski, Michael Kosfeld, Julia Lupp, Fred Thomas, Björn Vollan, Bruce Winterhalder, and Arnon Lotem as well as participants at the Ernst Strungmann Forum for valuable comments on an earlier version of this paper.

3

The Ecological and Economic Conditions of Exploitation Strategies

Maxwell N. Burton-Chellew, Alex Kacelnik, Michal Arbilly,
Miguel dos Santos, Kimberley J. Mathot, John M. McNamara,
Friederike Mengel, Joël van der Weele, and Björn Vollan

Abstract

In many situations across biology and economics, there is often one individual, or "agent," that invests effort into a beneficial task and also one individual that, in contrast, foregoes the effort of investing, and instead simply exploits the efforts of another. What makes an individual choose to invest in production versus exploiting the efforts of another? If everyone invests, then exploitative strategies become very profitable; however if everyone is exploitative, there will be no investments to exploit. How does natural selection resolve this dilemma? What can economic institutions do to encourage investment? Can biologists and economists learn from the approach of each other's discipline? This chapter outlines the commonalities and differences in approach of the two disciplines to the general problem of investment versus exploitation. It develops a model to encapsulate the general features of many scenarios ("games") involving potential exploitation and explores the benefits of a unified approach, outlining current limitations and important areas for future investigation.

Introduction

Picture one of your distant ancestors, eking out a living during the Paleolithic era, spending many hours carefully and patiently crafting some flint stone into a usable hand ax that he (or she) can use later to chop firewood, butcher meat, and dig out nourishing tubers. After several hours work, his axe is now ready, but just as he stands back to admire his handiwork, another man rushes by and takes the hand ax for himself. Your ancestor has spent valuable time and effort, using skills that took years to acquire, in crafting a usable tool, which is now only going to benefit another man—a man who may have spent the entire day resting or wooing potential partners, before exploiting the efforts of your industrious ancestor. In the evolutionary struggle for survival, it would appear that your ancestor is at a distinct disadvantage.

Alternatively, you may prefer to consider a New Caledonian crow, *Corvus moneduloides*, spending considerable time stripping and bending a twig into a tool fine enough to extract a nourishing but stubborn beetle grub hiding within the apparent safety of tree bark. After several minutes investigating the specific problem and the length of the local twigs, the crow then selects the most appropriate twig and shapes it accordingly. The twig is now ready to be used, to provide a beneficial return on the crow's investment of time and energy, but just as it digs out the recalcitrant grub another crow appears—a crow that had perhaps been resting or mating nearby—and helps itself to the tasty snack. Again, the more industrious individual appears to be at a disadvantage.

In both of these examples there is one individual that invests effort into a task with the aim of reaping a later benefit, and one individual that, in contrast, foregoes the effort of investing and instead simply exploits the efforts of another. Here there is a puzzle, for it would seem that the one that competes for instruments without toiling to make them is at an advantage to the one that produces them, for parasitizing upon the efforts of the producer saves time and energy. Surely, therefore, life's winners will adopt this parasitic strategy. However, if everyone is parasitic, there will be neither tools nor food to steal. So maybe everyone should choose to produce. Likewise in this case, if everyone is productive, a potential exploiter will have lots of potential victims and so being an exploiter will be more tempting. Our aims here are to explore the similarities between the biological and economic approaches to studying the above problem concerning investment versus exploitation and, more importantly, to investigate whether the empirical findings and theoretical concepts from one field can inform the other.

Investment versus Exploitation

The above examples show that individuals often face a decision between producing and taking, between relying on themselves to find and produce resources or relying on others to do so, between investing in production versus exploiting the investments of others. An adult individual can generally either invest in being self-sufficient, producing their own necessary resources, or they can adopt "parasitic" or "predatory" strategies that take advantage of, and thus "exploit," the efforts of others. Although the best terms to describe these alternatives are debatable, and although many individuals can and will also cooperate to find, produce, or trade resources, it is hard to argue that the decision to invest or exploit is not a recurring feature of life that shapes much of the biological world. For instance, many species predate on others, and it is possible that every multicellular species is at times exploited by multiple parasites, with the number of parasite species estimated to outnumber nonparasite species, perhaps by even 4 to 1. The presence of exploitative parasites has likely been a major driver of evolutionary change in host species (Price 1980; Moore 2002; Agosta et al. 2010).

Biologically, all organisms can be considered as either autotrophic or heterotrophic. Autotrophs, such as some plants and some microbes, obtain all the energy they need from their abiotic environment, deriving their energy from sunlight or inorganic chemicals. Heterotrophs, in contrast, rely upon the investments of autotrophs. Imagine, for example, a zebra that is grazing on the savannah: This zebra is a heterotroph that is exploiting the production of autotrophic grasses in a way that benefits the zebra but harms the grasses it consumes. Now imagine that this grazing zebra is suddenly attacked and eaten by a lion. This lion is likewise clearly exploiting the investments of the zebra, in a way that benefits itself, but harms the zebra.

The lion that eats the zebra that eats the grass is far from the end of the chain, for as explained above, both the lion and the zebra will likely be playing host to several small harmful parasitic species, which could be considered to be exploiting their host. Of course, such predation and parasitism are not the only available interactions. For example, the lion and the zebra will also be playing host to several microbiotic species that provide benefits as part of a cooperative mutualism. In fact, any and all actions that organisms perform (e.g., pigs depositing nitrogen-rich dung or earthworms moving nutrients through soil) may well provide a mix of unintended positive and negative effects to other species. However, the point remains that once there are investors, of any sort, opportunities for exploitative strategies follow. This exploitation is not restricted to predation and parasitism of the flesh. For example, lions will exploit the searching efforts of vultures by following them and monopolizing any carrion they find (Schaller 1972), and when the cuckoo lays its egg in the nest of another species, it does so to exploit the parental investment of another species, tricking them into caring for their own young, a trick referred to as "brood parasitism" (Davies et al. 2012).

When one species evolves to harm or exploit the investments of another, an evolutionary "arms race" typically occurs, where neither species gains a large relative advantage despite increasing adaptations for winning the conflict (Davies et al. 2012). This is because as one species evolves better defenses against exploitation, the other species evolves better countermeasures in response. For example, prey species will often evolve to run faster, and species that suffer from brood parasitism tend to evolve distinctive egg patterning which helps them identify and reject distinctive cuckoo eggs. However, as prey become faster and harder to catch, the selection pressure upon predators to become faster increases, and vice versa, creating the phenomenal speed of the cheetah. Both sides are running, but neither is getting ahead by much, in evolutionary terms. Likewise, as the victims of cuckoos make their eggs more distinctive, the selection pressure increases upon cuckoos to effectively mimic the egg patterning of their hosts (Dawkins and Krebs 1979; Davies et al. 2012). Again, both sides are adapting to each other, making life difficult for each other, but neither is ever winning by much.

Most of the above examples describe interspecific interactions, that is, interactions between species. However, the decision to invest or to exploit the efforts of others is central to many intraspecific interactions too, where members of the same species compete for the same resources, as in our opening human and crow examples. Arguably the most informative intraspecific example for biologists is in the foraging behaviors of gregarious bird species that move in flocks. As we explain more fully below, many birds face the choice to search for their own food (investment) or to save on energy and perhaps time by watching and exploiting the searching efforts of others before rushing in at the last moment to compete for a share of the bounty. Biologists have often termed these behaviors as "producing" and "scrounging," respectively, and considered these behaviors as separate strategies or tactics that individuals can employ while playing the "producer–scrounger game" (Barnard and Sibly 1981; Barnard 1984; Giraldeau and Caraco 2000; Giraldeau and Dubois 2008). In this case, to scrounge does not usually mean to beg or rely on the goodwill of producers but rather to compete for the food that has been discovered by another; hence we consider it to be behavior that exploits the investments of others. The producer–scrounger (PS) game has been used to conceptualize the range of foraging behaviors within many species of birds and mammals, including several primate species such as mangabeys, baboons, gorillas, and chimpanzees (see references within Arbilly et al. 2014). In general, we refer below to the behaviors in the PS game as an interaction between "investors" and "exploiters."

What is interesting in these intraspecific interactions is that rather than having two sides that evolve countermeasures to each other, we have similar individuals that can behave as either an investor ("producer") or an exploiter ("scrounger"). This means that the ratio of investors to exploiters and/or the decision rules employed to choose between acting as an investor or exploiter may change over time. These changes in behavior can theoretically occur either within individuals over time, through processes such as learning or plasticity, or within populations through genetic evolution. Biologists have therefore studied how various ecological and behavioral factors, such as the distribution of food or group size, affect the ratio of producers to exploiters in both wild and laboratory populations, and what information individuals (primarily birds) use to adjust their choice of strategy (Vickery et al. 1991; Giraldeau and Caraco 2000; Giraldeau and Dubois 2008).

that the customer will often be desperate for money to pay for primary needs (e.g., food, energy, heating) can make this black and white distinction a blurry gray. Second, the customer does not have the option to exploit the payday loan company. In this way, the asymmetry between the two agents means the situation more resembles an interspecies interaction than an intraspecies one (a zebra cannot become a lion). The effect of asymmetries between individual agents is a topic we explore below when considering how differences in factors, such as foraging ability and dominance, affect an individual's decision to invest or to exploit, but we restrict ourselves to examples where all individuals can, theoretically at least, choose between being an investor or exploiter in the relevant scenario.

The Biology of Individuals versus Groups

If a group of individuals all choose to exploit, there will be no production from which to draw an advantage, and such a group may be outcompeted by a group with more investors. It is therefore tempting to conclude that exploitation will not be favored. Such an explanation is, however, unsatisfactory because selection or competition does not only occur among groups but also within them. Therefore, within-group dynamics, where exploitation may provide a relative advantage, must also be considered (Williams 1966).

Darwin realized that natural selection favors the heritable components of physiology and behavior (the phenotype) that increase an individual's survival and reproduction. This process creates an appearance of design in organisms as they become increasingly well fitted (adapted) to their environment. This process of adaptation is why, within evolutionary biology, many organisms can be reasonably modeled as agents approximately maximizing their own survival and reproduction (Grafen 2006; Gardner et al. 2011). However, the heritable basis of an individual's phenotype is encoded by genes, and thus a complete understanding of evolution requires an appreciation of genetic success and survival.

Adopting a gene-centered approach to adaptation has allowed biologists to show that a gene is selected depending on both how the phenotype it encodes affects the bearer's reproductive success and how it affects the reproductive success of the bearer's relatives (Hamilton 1964). This is because a gene can increase its frequency within the population through two ways: a direct route whereby it produces a phenotype that helps its bearer to have more or better offspring and an indirect route whereby the phenotype helps other individuals (if they also contain identical copies of the gene) to produce more or better offspring. The simplest and most effective way to achieve this indirect route is by helping close relatives. However, often an adaptation that increases the success of one of these routes will come at the cost of reducing the success of the other. In these cases, selection will favor the optimal trade-off between the two, depending on the cost and benefits of helping, and the relatedness between the actor and the recipient of the behavioral act. This is because relatedness describes, for any gene that encodes behaviors with social consequences, its above-chance probability of affecting a copy of itself (the indirect route).

The evolutionary fitness of an individual's behavior can therefore be described or measured as the sum of their direct and indirect reproduction, referred to as "inclusive fitness." In the context of social interactions, this concept has allowed biologists to retain a legitimate sense of agency by adopting the perspective of individual actors trying to maximize their inclusive fitness (Hamilton 1964; West and Gardner 2013). Inclusive fitness theory has been very successful in identifying situations where individuals will be selected to reduce or increase their exploitation of others, depending upon their relatedness to them (Gardner et al. 2011). Evolutionary biologists have also demonstrated that models which focus on group-level benefits to explain cooperative behaviors still only get cooperation to evolve when the genetic costs and benefits of the behavior to the individual provide an inclusive fitness advantage (Lehmann et al. 2007; Marshall 2011). This means that natural selection will only favor individual behaviors that serve to benefit their group if they also serve to benefit either the individual or the individual's relatives (Gardner et al. 2011). There have, however, been some notable disagreements with this view (Nowak et al. 2010; Allen and Nowak 2015), especially with regard to human evolution (Henrich 2004).

Modeling contemporary human behavior is not so straightforward, largely because of two factors. First, because our environment has undergone rapid change, we may no longer be as well adapted to our environment as we once were. For example, there is an evolutionary lag between our cravings for energy-rich fats and sugars and the modern glut of such foods in certain countries. Sugary and fatty foods were highly beneficial, but rare, in our ancestral environment, so the costly consequences of such cravings were minimal, unlike today where they cause obesity and heart disease. It is likely that the change in our available diet is not the only domain where we are lagging behind. Other potential areas are the sexual domain and our relationship with pornography and reproductive technologies, such as contraception and abortion. Likewise, it is highly unlikely that we are well adapted to handle money and the complexity and longevity of modern financial products, such as pensions.

Second, the human capacities for problem solving, learning, forward planning, and communication indicate that we are, in principle at least, capable of organizing collective behaviors that benefit society. There is debate about how we should model the evolution of group-beneficial behaviors in humans, and whether the fundamental predictions of evolutionary theory still apply (Henrich 2004). If our social traits are transmitted through cultural routes, such as imitating others, more than through genetic routes, then cultural evolutionary models will be needed to explain and predict the distribution of social behaviors (Henrich 2004; Richerson and Boyd 2005). Contrary to the predictions of genetic models, the outcomes of cultural models are less certain and depend largely on the precise nature of any genetically based biases we exhibit in learning and copying (Aoki et al. 2011; Mesoudi 2011; El Mouden et al. 2014; Acerbi and Mesoudi 2015).

The Economics of Individuals versus Groups

The situations we discuss in this chapter often involve individuals making a choice between what is best for them and what is best for the group. For example, if more people invested in giving blood or contributing to Wikipedia, then society would be better off, but this means that individuals would have to give up some of their time. If more parents had their children vaccinated, then the overall immunity of the society would increase, but parents may feel that vaccinations expose their children to undue risks. If individuals falsely report their earnings, they will pay less tax, but then society has less revenue to use for communal services. If fishermen catch fewer fish, they earn less money, but the shared stocks have a greater chance of replenishing themselves. In all these cases, the interests of the individuals making the decisions diverge from the interests of the larger group, or society, in general.

How then should such decisions be modeled? Traditional economic models of decision making have often assumed that people know what they want and make their choices accordingly. More specifically, traditional models assume that people consider the consequences of their choices, rank the possible outcomes by how desirable they are, and then choose an action that leads to their most preferred outcome, or act as if they were doing so. Such models have the advantage of not requiring the modeler to assume to know what people want. Instead, one can infer an individual's preferences from their choices (the principle of revealed preferences). This principle does not allow us to question whether someone is making a decision in line with their preferences or not, because of the assumption that individuals prefer the outcome they choose, and choose the outcome they prefer (Kreps 1988; Kacelnik 2006).

The above principle of revealed preferences relies on *rational choice theory*, which assumes that people make choices consistent with their preferences. This assumption can only be tested by examining the pattern of choices individuals make (Allingham 2002). More formally, given an individual's preference ordering, which satisfies properties of consistency and completeness (the preference ordering covers all permutations of options), this model can be conveniently formulated as the maximization of a utility function. In this general framework, *utility* is merely a numerical ordering of outcomes; it does not make assumptions on the specific preferences of an individual other than the consistency and completeness requirements.

While rational choice does not imply that people cannot care about others' outcomes, it has often been understood to mean that people prefer to maximize their own monetary and/or material gains. This assumption greatly simplifies models of decision making and has proved to be a reasonable approximation in many economic domains, such as market competition. This assumption, however, has not fared so well in social domains, where people have to make decisions that affect both themselves and others. Hundreds if not thousands of experiments have repeatedly shown that people in laboratory conditions will often make decisions that are suboptimal for their own finances but beneficial for the finances of others, suggesting that people do not only value their own welfare (Chaudhuri 2011; Engel 2011; Guth and Kocher 2014).

Can groups be modeled the same way? If individuals make rational choices, then surely groups of rational individuals will do so as well. However, as Mancur Olson has shown, it is illogical to consider groups to be rational just because they are comprised of rational individuals (Olson 1965). Olson's logic is that if the individuals within a group are rational, they will follow their own interests; thus the group will collectively appear irrational whenever individual interests conflict with group interests. In other words, groups may well be rational, not because they are made up of rational individuals, but because group and individual interests are aligned, which is rarely the case. This parallels the lesson from inclusive fitness theory in evolutionary biology: individual behaviors only evolve to serve the group when they also provide sufficient benefits to the individual or to the individual's relatives.

Much research has therefore focused on the idea of increasing "efficiency" in scenarios where individual and group interests are in conflict. Efficiency concepts in economics relate to the aggregation and favorable trading off of the utility of different people. Such trade-offs are central to the discipline, as evidenced by the pervasive applications of cost-benefit analysis in policy applications. One problem is that standard utility concepts are based on preference rankings; they only provide (ordinal) rankings over alternatives. Thus, while (numerical) utility values describe the relative valuations for a given individual, the absolute level can be arbitrarily rescaled, making interpersonal utility comparisons impossible. To overcome this problem, economists have developed several different efficiency concepts to help theoreticians and policy makers compare outcomes.

The most commonly accepted notion of efficiency is "Pareto efficiency," because this concept obviates the need for interpersonal comparisons of outcomes. Instead, Pareto efficiency says that an increase in efficiency, a Pareto improvement, occurs when a new allocation of resources makes at least one individual better off while keeping every other individual at least as well off. When no Pareto improvements are possible, the outcome distribution is Pareto efficient. Note that Pareto efficiency does not imply equality; it is perfectly possible for a distribution of resources to be unequal and Pareto efficient, as long as any changes toward greater equality leave at least one person worse off.

Thus, many economic analyses focus on "correcting" deviations from (Pareto) efficiency that would result from individuals pursuing their own interests. This can be accomplished through various institutions designed to shift the relative payoffs of different choices for individuals. In this way, they attempt to incentivize more desirable behavior in rational individuals (Hurwicz and Reiter 2006). Examples include monetary incentives such as taxes, subsidies, and fines but also manipulations of social incentives such as prestige, shame, and guilt (Kosfeld and Neckermann 2011).

Furthermore, people often appear motivated by a concern for the welfare of others (other-regarding preferences) that may act to reduce the conflict between the individual and the group (Fehr and Schmidt 1999; Bolton and Ockenfels 2000). Thus, efficiency analysis may be improved by including people's sense of fairness when considering trade-offs between economic growth and inequality (Durante et al. 2014). It is also possible that traditional policies have been too pessimistic in assuming people's preferences are entirely self-interested. By making salient the idea that people are selfish, such policies may have provided a justification for selfish behavior and reduced people's faith in each other. If such policies are self-fulfilling, it may instead be better to focus on increasing people's confidence and trust in each other (Gaechter 2007; Bowles and Hwang 2008).

Given all of the above, how then can we analyze the situations of investors and exploiters in which we are interested? It would seem that we have to consider the costs and benefits of the decision to invest or to exploit at the individual and not the group level. However, if people are rational and self-interested, then why will there ever be any investors? If exploiters gain a higher payoff, then surely everyone will choose to exploit?

Let us return to our original examples of one of our ancestors laboring to make a hand ax that ultimately benefitted another man, and a crow laboring to extract a grub that ultimately benefitted another crow. In these extreme cases, there is no benefit to the individual producer, but imagine that instead of the ax being stolen straight away, our ancestor was able to use the ax for a day, or a week, or a year, before it was stolen. In this case, the costs of making the ax would have been recouped, provided that the benefits were large enough to have outweighed the costs. Likewise, imagine that the crow, instead of extracting one beetle grub, had extracted several and gotten to eat some of them before a competitor came and took some for herself. Here, the investments of the industrious crow returned a "finder's share" or "advantage" (Vickery et al. 1991). More generally, if the average finder's share sufficiently outweighed the costs, then we could argue that it is sensible to sometimes be an investor. In human societies, creating or enlarging such a finder's share is the aim of regulations like patent law, to which we will return to below. The effect of a finder's share in regulating fisheries is also discussed by Valone et al. (this volume).

Of course, even with a large finder's share, it will still sometimes be sensible to be an exploiter, for as we have outlined, whenever there are investors, there are opportunities and benefits for exploiters. Therefore, the predicted ratio of investors to exploiters will depend on the circumstances, the behavior of others, and the associated costs and benefits of the different behaviors, which will depend on how costly it is to invest in a resource, how many potential investors there are to exploit, the cost of exploiting in terms of defense or retaliation by investors, and the value of the resource being produced (Barnard 1984). Calculating exactly how the incentives, and therefore the ratio of investors to exploiters, will change requires mathematical models.

Applying the Concepts of Investment and Exploitation to Biology

In biology, examples of investor–exploiter relationships are widespread across all scales of life. For example, in bacteria, some individuals invest in production while others exploit their efforts. Here, investors make and release siderophores, molecules which sequester valuable and metabolically essential iron from the local environment (Ratledge and Dover 2000). Siderophores, however, are costly to produce and are not guaranteed to return to the original investor. Therefore, other bacterial strains in the same neighborhood can still survive even if they do not invest in siderophore production by harvesting the siderophores produced by others (West and Buckling 2003). These microbial examples are useful because we expect them to conform to the laws of the relentless evolutionary process. Furthermore, they show that cognition and intentions are not necessary to replicate investor–exploiter dynamics. Microbial examples can thus be used to test the predictions of theoretical models without having to worry about the role of intentions or the effects of either imperfect or sophisticated cognition.

Investors do not only produce food and metabolites. The various strategies that animals use to obtain sexual partners can also be viewed as alternative mating tactics within an investor–exploiter framework. In many frog species, some males will invest in calling to attract females, putting themselves at increased risk of predation, while other males will exploit these investments, thus avoiding the increased predation risk, by staying silent, and intercepting females as they approach calling males (Lucas et al. 1996). Likewise, brood parasitism is another major class of reproductive tactics that can be effectively categorized into investor and exploiter roles. In many species of birds, females will surreptitiously lay their eggs in the nest of another female, exploiting the incubation and provisioning efforts of the nest-attending female (Davies et al. 2012).

Arguably the best-studied example of the use of investor–exploiter tactics comes from the field of social foraging (Afshar et al. 2015). When animals forage in groups, individuals can invest in searching for food, or they can exploit the food discoveries of others. This investor–exploiter scenario has been extensively studied using models that typically can only be applied to groups of more than two individuals and contain simplifying assumptions that may not be true of all such scenarios. Typically, researchers make three assumptions in these investor–exploiter models (termed PS models in much of the literature):

1. Tactics are mutually exclusive; that is, individuals cannot simultane-
    ously search for resources and search for opportunities to exploit the
    resources produced by others. For example, in certain species of fish,
    males can either invest in building/maintaining a nest to attract females
    or they can try to exploit the nest building of others and sneak paternity
    as females approach another male's nest. By definition, a male cannot
    do both of these things simultaneously, although of course they can
    invest in doing a bit of both over time, but this means they have to al-
    locate each unit of effort into investment or exploitation. Similarly, in
    many birds, the nature of their visual system means that searching for
    food patches on the ground requires a different head orientation than
    searching for congeners that have discovered food patches (Coolen et
    al. 2001).

2. The resource is finite and can be depleted, so that any part of the re-
    source consumed or used by one individual is not available for other
    individuals. For example, food that is consumed by one individual is
    not available to others, or eggs fertilized by one male can no longer
    be fertilized by another male. In economics, as explained below, such
    goods are considered "rivalrous," because different individuals can be
    considered as rivals competing for the same goods.

3. Investors, on average, gain some advantage for having produced a re-
    source. For example, they get to eat some amount of food they discov-
    er before another individual joins them. This is known as the finder's
    share (discussed above).

Several empirical studies have tested these assumptions (Giraldeau and Dubois
2008; Dubois, this volume; Barta, this volume). The predicted effects of group
size, patch size, and finder's advantage on the frequency of exploitative behaviors have been experimentally tested in laboratory experiments with nutmeg
mannikins (*Lonchura punctulata*), a small ground-feeding passerine, and the
results qualitatively support the predictions. For example, doubling flock size
from three to six individuals resulted in a 50% increase in the frequency of
exploitation (Coolen 2002). Increasing the number of seeds available per food
patch also resulted in increased exploitation (Coolen et al. 2001; Coolen 2002).
In a very elegant experiment, Mottley and Giraldeau (2000) experimentally
manipulated the finder's advantage by devising an apparatus where individuals had to pull on a string to access seeds in a food patch. The seeds then fell
into a collecting dish, and the finder's advantage was manipulated by partially
covering the dish in such a way so as to limit the number of seeds that could
be consumed by the producer. As predicted, when the finder's advantage was
reduced, the frequency of exploitation increased.

Field experiments also provide support for the main predictions of investor–
exploiter models. Experimenters allowed free-living Carib grackles (*Quiscalus
lugubris*) to invest effort in making dry pieces of dog food more palatable by

dunking them in puddles, or to attempt to exploit the efforts of others by stealing food from group members after it has been softened. Field observations by
Morand-Ferron et al. (2007) have shown that the proportion of exploitation attempts changes with natural variation in wild Carib grackles group size (more
exploitation in larger groups). In their study, Morand-Ferron et al. also experimentally manipulated the expected finder's advantage for investing in dunking
dog food. To do this, they created experimental puddles of equal surface area
but varying in shape, which changed the average distance to exploitation opportunities. Larger distances to exploitation opportunities meant that investors
were more likely to consume the food before an exploiter could attempt to
steal it (i.e., their average finder's share increased). As a result, the frequency
of exploitation declined under these experimental conditions.

The above experiments show that natural selection does not have to act upon
genes for being an investor or exploiter, but instead can select for behavioral
changes based upon genetically encoded learning rules. If investor–exploiter
scenarios are sufficiently common and important, then animals will evolve
mechanisms to improve their dynamic performance in repeated instances of
such "games."

Applying the Concepts of Investment and Exploitation to the Social Sciences

The social sciences often study the conflict between individual and group
interests by examining either investment in the production of goods that are
freely available and benefit society (public goods), or the private consumption
of shared resources (common-pool resources). In these scenarios, individuals
can pay a cost (invest) to produce directly a good that benefits everyone, or
they can show restraint in their consumption and "invest" in future resources
by avoiding overconsumption of a public resource. In both cases, individuals
typically benefit more by acting against the interest of the group, and thus
outcomes are predicted to be suboptimal at the group level. The scale of this
individual versus group dynamic can vary: sometimes the "group" is one local
group, or a country, or even the global population, and the conflicting "individual" could be a single person, a region, or even a country, respectively. In all
cases, however, the "individual" is a subunit of the larger "group."

Various mechanisms have been proposed to increase the group welfare
when individual- and group-level interests are not aligned. Not all scenarios
are, however, the same, and because the efficacy of different mechanisms will
depend on the nature of the goods being "produced," the social sciences have
found it useful to categorize public goods into different types.

Goods can be conceptualized along two dimensions, according to their degree of "rivalry" and "excludability." Rivalrous goods are those which can be
used by only one person simultaneously (e.g., a candy bar or a chair), over
which people may wish to compete. In contrast, nonrivalrous goods are those

that can be used by many at the same time, and are therefore often intangible (e.g., national defense or scientific knowledge). Excludable goods are those for which one can limit the benefits to only those that contribute to the costs. For instance, one can exclude noncontributors from a mail service by requiring people to pay for a stamp. In contrast, one cannot prevent noncontributors from using or benefiting from nonexcludable goods. For example, it is not easy to prevent people from listening to a radio or television broadcast, or restrict access to many natural resources (e.g., fish or timber stocks).

Taken to their extremes, the combination of rivalry/nonrivalry and excludability/nonexcludability allows one to distinguish between four types of goods. In no particular order, there are private goods, which are often provided commercially, whereby people pay to use/consume an excludable and rivalrous good, such as a sandwich: here production and consumption are in the self-interests of both parties. Second, there are club goods, which are those that are excludable but effectively nonrivalrous because one additional user does not significantly affect overall costs, although too many users may lead to congestion. For example, users of a golf club can exclude nonmembers and benefit from the membership fees of each additional member, but may want to limit the number of members to avoid overcrowding at busy times. Third, there are public goods, whereby society pays for goods that are freely available to everyone (nonexcludable and nonrivalrous). Public goods are expected to be underprovisioned because self-interested individuals will be reluctant to invest to provide public benefits. Finally, there are common goods, also referred to as common-pool resources, which like public goods are also nonexcludable but, unlike public goods, are rivalrous. Fishing grounds provide a good example, because it is difficult to exclude boats from fishing, but any fish caught by one boat reduces the catch available for others. Therefore a recurring problem with common goods is how to prevent overuse. Rational individuals are predicted to maximize consumption because any restraint they show can be exploited by others increasing their consumption. Thus, well-defined property rights for common goods can theoretically solve the problem of overuse but can be costly to monitor and enforce, and may be unpopular.

How does the PS game relate to this classification of goods? If we imagine a social forager that chooses to invest in searching for food or to exploit the searching efforts of others, then it is apparent that the food produced is clearly rivalrous, but not fully excludable, if at all. The finder's share sits in the continuum between full and nonexcludability; thus, this classification of goods may not be the most useful to use in framing the PS game. Instead, the idea of rent seeking from economics may be a better fit to the PS game. Rent seeking is a broad concept used to describe efforts devoted to the capture of resources by diverting them away from other agents (Tullock 1967, 1974; Murphy et al. 1993; Congleton et al. 2008). While the definition is somewhat fluid, it crucially refers to actions that, in contrast to this chapter's use of the term production, do not create new value or surplus. The term rent seeking originates with

Adam's Smith's classification of income into profit, wages, or rent, and refers to the attempt to gain control of (natural) resources.

Gordon Tullock designed a flexible game theoretic framework where participants exert efforts to compete for the possession of a resource, called the Tullock contest or rent-seeking game introduced above (see also King et al., this volume). This framework has been applied to a range of activities, such as lobbying for government protection via restrictions to foreign imports or limiting access to lucrative industries like notaries and pharmacies. Dechenaux et al. (2015) review experimental literature that has studied behavior in this game in the laboratory, and Murphy et al. (1993) discuss an application to economic growth and development, where rent seekers target surplus-generating activities like innovative companies or high-yield farming. The authors distinguish between private and public rent seeking. The former consists of theft or piracy, whereas the latter includes government corruption in the issuance of licenses or other government services. The authors suggest that innovative activities are especially vulnerable to public rent seeking, as innovators are dependent on government services and do not have established lobbies to obtain administrative favors for themselves. As a consequence many companies prefer to operate in the informal sector (and not become too big) and to operate under the radar of the corrupt government.

Murphy et al. (1993) include a discussion of the frequency dependence of the returns to productive and rent-seeking activities. As one may expect, the return to productive activities causes a decline in the amount of rent seekers in the economy. However, the authors propose that there are complementarities in rent seeking which may cause the returns to increase with the amount of rent seekers, at least for low levels of rent seeking. Sources of complementarities may be strength in numbers among rent seekers, as embodied by gangs or mafias, or the escalation of arms races in the fight over resources. Depending on the exact nature of the frequency dependence, multiple equilibria may exist that are characterized by either high levels of rent seeking and low levels of production or vice versa. Murphy et al. (1993) suggest that such multiplicity may help explain variances in the development of different countries.

One application of the rent-seeking model in the context of developing countries is to the phenomenon of forced interfamily solidarity, where the requirement to share within the extended family network is an integral part of everyday life, one exacerbated by the lack of formal insurance systems (this was first reported by anthropologists in the 1960s; Belshaw 1965). The literature on "forced solidarity" or the "dark side of social capital" shows that acts of giving are not always voluntary and are often "demanded" from the network members, thereby deterring investment and effort of the producer (di Falco and Bulte 2011). Entrepreneurs must resist normative pressures to support their extended families if they wish to reinvest in their firms (Belshaw 1965; Nafziger 1969; Bloch 1973; Hart 1975). This has led to the hypothesis that high demand for financial support by members of an entrepreneur's family hinders savings and investments (in capital and labor) and thus may hinder long-term economic growth. The above similarities between the concepts and models of the PS game and rent seeking are another example of parallel developments in the fields of evolutionary biology and economics. Both fields utilize a cost-benefit approach with the concept of equilibrium thinking, but alas their use of different terms for similar phenomena can create a potential barrier to interdisciplinary collaboration between like-minded researchers.

There are scenarios that are still not captured by either investment in public goods, the PS game, or the rent-seeking game. These include situations where the investments of several members are required to produce a good, and situations where the investments of one or a minority of members are sufficient to benefit all members of a group. Here, the benefits of investment are not directly proportional to the number of investors. Imagine a group of lions hunting for a buffalo. More than one individual lion is required to kill the buffalo, so a team of lions is required, but each lion is better off letting other lions take the risk of attacking the buffalo. A modern human example is when people profit from open-source software. More than one producer is required to generate it, but everybody has access to the software. Alternatively, if several people witness an accident or crime, it only takes one person to phone the emergency services; this "diffusion of responsibility" can create the possibility that everyone leaves the task to someone else (Darley and Latane 1968). Situations requiring a certain threshold of investors before a potential good can be realized have been studied in economics as threshold public goods games, and there is a considerable experimental literature on such games (Croson and Marks 2000).

It has been argued that this type of scenario, the so-called volunteer's dilemma (Diekmann 1985, 1993), also known as the snowdrift game, chicken game, or hawk–dove game (Maynard-Smith and Price 1973; Maynard-Smith 1982), is more relevant to the study of exploitation, especially within biology, than the classical linear public goods game (Archetti and Scheuring 2011). As in the PS game, there can be an equilibrium point between strategies that invest and exploitative strategies in the volunteer's dilemma. The idea of requiring a threshold number of investors to produce the actual good is a potential extension of the PS game for future research. It is reasonable to assume that one investor might not be sufficient to produce any goods. Alternatively, the presence of more than one investor at a patch (e.g., by chance, in a world where the number of patches is finite) might be required to attract exploiters. Such modification would probably decrease overall resource use because all lonely investors would not be able to profit from the resource (or attract exploiters), which would decrease the population's average payoff.

Finally, there has sometimes been concern expressed that if individuals receive help from willing investors, they are then less likely to become investors themselves. This has been termed the Samaritan's dilemma, whereby an altruistically motivated person has to decide between leaving a helpless person to suffer and providing help at the risk of reducing the recipient's self-reliance

(Buchanan 1975). If a potential recipient knows that the Samaritan will help, then his/her incentive to work may, theoretically speaking, be reduced. Such arguments can be used as a rationale for reducing publicly funded welfare or social security for people who face unemployment. However, such an argument may rely on the benefits of employment being less than those of unemployment, unless one is arguing that some people are of a different disposition to others with regard to their desire to work. A meaningful analysis of such a dilemma requires more understanding of the psychological and sociological forces that impact an individual's ability, desire, and opportunities to work.

Model Behavior

If decisions are based upon the costs and benefits of different outcomes, then such decisions have to be modeled mathematically. Sometimes a necessary trade-off of such modeling is that complex real-world phenomena have to be studied with simple "games" that abstract away much of the important complexity. This can generate valid concern that the subsequent analyses are overly simplistic and reductionist. However, a cost-benefit approach is necessary to model evolutionary outcomes because the process of natural selection blindly operates on differential success in terms of reproduction. In economics, as outlined above, the cost-benefit approach can be justified by the conservative expectation that people prefer to have more resources/money. While this assumption can be falsified in certain cases, it vastly simplifies the mathematical models and makes the analysis feasible. If we relax this assumption, but maintain the idea that individuals maximize something, as defined by their own personal utility function, then whatever is being maximized can be mathematically modeled (Kacelnik 2006).

Sometimes the best behavior or "strategy" depends on what others are doing. For example, the costs and benefits to a crow that uses a tool to extract grubs depend on how many other nearby crows are being exploiters. This decision can be thought of as stemming from a game, whereby it would be beneficial for an individual to know what other individuals are doing before making a decision on how to respond strategically. One way to model such situations is through game theory, which can be used to calculate what the stable outcome of such games will be, provided the "players" can either anticipate the moves of other players or change their strategy accordingly in response to success or failure. The adjustment of strategy choice can be through natural selection acting upon the differential success of competing versions of the same gene (alleles), through various economic processes that favor organizations using more successful strategies, or through various learning processes within individuals and organizations. However, if the success of a strategy relies upon it being rare, then as successful strategies become more common they will enjoy increasingly less success, which will limit their growth in the population. This dynamic is known as negative frequency dependence (Figure 3.1).

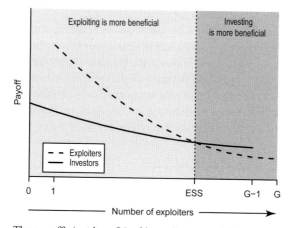

Figure 3.1   The payoffs (net benefit) of investing or exploiting are frequency dependent. When there are many investors, exploiters obtain a larger payoff than investors because they invest no effort in discovering resources, and there are a large number of investors to exploit. In contrast, when there are many exploiters, the average payoff of exploiting decreases because investors are harder to find and there is more competition to exploit them. At some intermediate mixture of investors and exploiters, both tactics receive equal payoffs (dashed vertical line). This is known as the stable equilibrium frequency, and no one individual can use a shift in tactic to increase their own payoff. ESS: evolutionarily stable strategy.

One crucial concept used to predict the outcome of negative frequency dependence, in both biology and economics, is the idea of equilibrium, which can be thought of as the solution to, or the predicted outcome of, the above types of game. This is the idea that when a population of individuals reaches a certain state, whereby there is a certain proportion of individuals adopting a certain behavioral strategy, no one individual can improve their situation by changing their strategy. This is key for economists because it means that no rational individual will be predicted to change their behavior unless they want to incur a loss, and that other rational individuals can use this assumption to anticipate the behavior of other players, leading to a stable outcome if players are rational and have correct beliefs; this is known as a Nash equilibrium (Nash 1950). For biologists, the concept is equally crucial, because it identifies the evolutionary resting state of the population where individuals will behave according to an evolutionarily stable strategy (ESS) (Maynard-Smith and Price 1973; Maynard-Smith 1982).

A strategy is evolutionarily stable when no plausible mutant strategy can improve upon it. The process also applies to the short-term changes in animal behavior when animals use learning to respond to changes in their own payoffs and/or in the behavior of others. We therefore expect animals to be well adapted to their social environment, even if we observe different behaviors in different individuals of the same population, or different behaviors over time within the same individual. Thus a game theoretical mathematical approach, while limited, offers potentially large benefits to the analysis of social behaviors such as investment and exploitation. One may wonder if it is too hard for human rationality to compete with the rationality of the evolutionary process (Kacelnik 2006). Nash himself, however, provided two interpretations of his equilibrium concept: one based on rational individuals that calculate the best action, and the other based on a large population of individuals in "mass action" that use their experience and limited knowledge to gravitate toward the equilibrium (Kuhn et al. 1996).

A Mathematical Model of Exploitation

Here we outline a mathematical model of exploitation that will form the basis of our subsequent analyses. Consider G group members engaged in joint production. The basic situation we are interested in is one where individuals can either invest in a productive activity or in an exploitative activity that parasitizes the investments and efforts of producers, such as in the PS game (Giraldeau and Caraco 2000).

We refer to the amount invested by an individual i into the productive activity x_i and into exploitation y_i. In the PS game, individuals either invest all their resources into production or into exploitation: $x \in [0, 1]$; $y = 1 - x$.

If an individual invests into production, the individual produces productive effort for T periods and produces an amount F with success rate λ. Parameters that affect who benefits from production, β and δ, allow a direct comparison between the PS and public good models. If a proportion $(1 - p)$ of the individuals in the group produces, then the payoff (fitness) of any individual producer $(x = 1)$ is:

$$W_x(p) = [1 - (1 - \beta)pG]T\lambda\left(a + \frac{F - \beta a}{\delta G + \beta}\right).$$ (3.1)

The payoff of an individual using the exploitation strategy $(y = 1)$ is:

$$W_y(p) = (1 - p)GT\lambda\left(\frac{F - \beta a}{\delta G + \beta}\right).$$ (3.2)

Under the PS model, $a > 0$ represents the "finder's share" discussed above in our crow example, a small reward to the producer for being the first to enjoy the benefits of production. Further, $\beta = 1$ and $\delta = p$; that is, the results of production are shared among all the exploiters as well as the producer of F, but not the other producers. While an exploiter in the PS model can benefit from the investments of all other group members, producers only reap the benefits of their own investments.

The Nash equilibrium of the game, where both strategies coexist and have equal payoffs, can be determined by equating Equation 3.1 and Equation 3.2, and solving for p. This yields the equilibrium frequency of scroungers $\hat{p}$, which is also the ESS since for any p above or below $\hat{p}$, the direction of selection will go toward $\hat{p}$; that is, $W_y(p) > W_x(p)$ if $p < \hat{p}$ and $W_x(p) > W_y(p)$ if $p > \hat{p}$ (Giraldeau and Caraco 2000):

$$\hat{p} = 1 - \frac{a}{F} - \frac{1}{G}. \tag{3.3}$$

In economics, socioeconomic interactions similar to the PS setting are typically modeled via rent-seeking models (Tullock 1967). In rent seeking, individuals can allocate any amount $x \in [0, 1]$ to production, while again $y = 1 - x$. The payoffs (fitness) of any individual i is then given by:

$$W(x, y) = \frac{y_i}{\sum_j y_j} \sum_j x_j, \tag{3.4}$$

where the sum of the productive efforts of all individuals gives the overall amount shared among all and the share each individual obtains is linearly proportional to the amount invested into exploitation.

The equilibrium amount each individual allocates to exploitation in the unique symmetric equilibrium (there are asymmetric ones as well) is:

$$\hat{y} = \frac{G(G-1)-1}{G(G-1)}. \tag{3.5}$$

This is only well defined for groups of at least size 2, for which individuals devote exactly half their resources to production. For larger groups, the amount devoted to production (x) decreases, just as it does in the PS model.

A second, and more common way to model the organization of trade-off between private and public interests in the social sciences is the so-called public goods game. Here, producers make personally costly investments into products, resources, or services that benefit the entire group. Typically these group-level benefits are assumed to be larger than the individual costs but not so much as to outweigh the individual costs of production. Therefore, the public goods game is different to the PS game in that $a < 0$: producers do not get a net reward for producing, but instead have to pay a net cost for the effort of producing. A second difference is that $\beta = 0$ and $\delta = 1$: all individuals in a group share the benefits of production equally, regardless of whether they are a producer or not. In other words, the costs of production are entirely private, but the benefits are entirely public. Thus, the income-maximizing strategy is generally to not produce, unless the benefits of production are sufficiently large for a given group size. As a result, under commonly assumed parameters, the unique ESS is $\hat{p} = 1$.

Here we outline a formal representation of this game that shows the similarities and differences to the PS paradigm and notation. In behavioral economic experiments, linear public goods games are implemented slightly differently and the customary payoff function is:

$$W_i(x,y) = \frac{\alpha}{G} \sum_j x_j + y_i, \qquad (3.6)$$

where individuals can allocate any amount x between 0 and 1 to production, while again $y = 1 - x$. The parameter α is an efficiency factor that determines how large the group-level benefits are, and is often chosen as 1.6 or 2.0 in economic experiments using groups of around four or five individuals (Isaac et al. 1984; Andreoni 1995; Fehr and Gaechter 2000; Fischbacher and Gaechter 2010). As long as α is not greater than G, the unique equilibrium is again $y = 1$. For example, in a group of four players with $\alpha = 1.6$, each dollar invested in production returns 1.6 dollars to the group, meaning that each individual gains 0.4 dollars and the contributor makes a net loss of 0.6 dollars.

Model Extensions

Models are advantageous in that they provide tractable, analyzable, but simplified versions of the real world. However, the wrong model, or an overly simply model, may not only be of limited use, it may actually provide results that are at odds with those from a more complicated model. Thus it is necessary to push for richer game theoretic models that incorporate more of the actual behavior involved (McNamara 2013). Here we consider the effects of several possible extensions to the PS game modeled above.

Limitations of the Game Theoretic Framework

Simple game theoretic models run the risk of deducing universal solutions that may not fit the problem at hand. Thus, defining ecological and socio-economic conditions under which more or less scrounging emerges is key to understanding systems dynamics, and potentially increasing the efficiency of these systems. The social-ecological system (SES) framework, proposed by Elinor Ostrom (2007) in the context of common-pool resources management, is a comprehensive way of describing the major variables that need to be accounted for in more realistic interactions between (human) agents in a given environment. The SES framework is used to explain the management of a resource according to rules and procedures determined by a governance system. On the most fundamental level, Ostrom distinguished between four fundamental properties of social ecological systems: the resource system, the resource units, the governance system, and the users of that system. Within these "first-tier" categories, specific "second-tier" variables are listed that have been tested empirically to influence the stability of social-ecological systems (Table 3.1).

Laboratory or field evidence of investment and exploitation in the PS game comes from very specific situations, which, of course, limit the ability to generalize these findings. The most common experimental setup involves socially foraging birds as the users competing for immobile resources. The interaction

Table 3.1 Variables in the social-ecological system (SES) framework, based on Ostrom (2007).

Resource System	Governance System
Sector (water, forests, pastures, fish)	Government organizations
Clarity of system boundaries	Nongovernment organizations
Size of resource system	Network structure
Human-constructed facilities	Property rights systems
Productivity of system	Operational rules
Equilibrium properties	Collective choice rules
Predictability of system dynamics	Constitutional rules
Storage characteristics	Monitoring and sanctioning processes
Location	
Resource Units	**Users**
Resource unit mobility	Number of users
Growth or replacement rate	Socioeconomic attributes of users
Interaction among resource units	History of use
Economic value	Location
Size	Leadership/entrepreneurship
Distinctive markings	Norms/social capital
Spatial and temporal distribution	Knowledge of SES/mental models
	Dependence on resource
	Technology used
Interactions	**Outcomes**
Harvesting levels of diverse users	Social performance measures (e.g., efficiency, equity, accountability)
Information sharing among users	
Deliberation processes	Ecological performance measures (e.g., overharvested, resilience, diversity)
Conflicts among users	
Investment activities	Externalities to others SESs
Lobbying activities	

among agents is limited to investing in producing food or exploiting the investments of others. Adding to this setup, behavioral ecologists have studied the variable influence of different factors, including the number of users, the size of the finder's share, the resource value and group size. Below we aim to highlight some of the evidence related to these second-tier variables or make projections on how they might shift the fraction of investors and exploiters in a given population.

Resource Heterogeneity

The basic PS game does not incorporate complex ecological conditions, such as variable amounts of resources and how they are distributed. We illustrated earlier that exploitative rent seeking might explain low growth in developing countries. However, some countries may be more prone to rent seeking because of the amount or type of resources they have. Sachs and Warner (2001) discuss that rent seeking and corruption may be more likely in countries that are rich in natural resources because such resources are more concentrated and appropriable by government officials. Thus, the productivity of the resource system, the predictability of the resource dynamics, and the temporal-spatial distribution of resources, to name just three of the key factors mentioned by Ostrom, might influence the amount of exploitation in a society. Here we aim to present some predictions on the effect of different resource characteristics on the proportion of exploitation.

Resource heterogeneity can arise when patches vary in the amount of resource they contain (i.e., when their quality varies). Recent simulations by Afshar and Giraldeau (2014) predicted that varying patch quality results in an increase in the frequency of exploiters in the population, a result they also confirmed empirically (Afshar et al. 2015). A potential extension of their model is to vary both the mean and variance in patch richness between groups and to allow for (potentially costly) migration between them, so as to determine their effects on the overall frequency of exploiters in the population. Another interesting extension of the baseline model of the PS game is to add a dynamical component to the resource. Barta and Giraldeau (2001) investigate a situation in which food patches are ephemeral, such as flying insects. In their model, once an investor has found a resource patch, both investors and exploiters do not have the time to deplete it completely. Under such conditions, the presence of exploiters might not decrease the investors' intake and resource use might not depend on the frequency of exploiters, because individuals do not have the time to compete for the entire resource and can only consume a fraction of it. Resource heterogeneity might be introduced in this model by varying the time available to consume the resource between groups and to allow for different efficacies of resource consumption (or searching efficacies) between individuals. In this case, individuals who are efficient in collecting the resource might be selected for, which would increase the population-level resource use.

In reality, resource patches are spatially structured in a finite three-dimensional space. Hence, the distance between resource patches is expected to have considerable effects on the population's average intake (Beauchamp 2008). Large distances between patches are likely to result in both lower encounter rates for investors (i.e., lambda in our mathematical model), but also lower detection rates for exploiters. Both theoretical and empirical studies have shown that lower probabilities of finding patches result in lower frequencies of exploitation in the population (Beauchamp 2008; Afshar and Giraldeau 2014). The rate at which exploiters detect or join investors will impact directly on the finder's share: if it is low, investors will have more time to profit alone from the patch, or will share the remaining resources with fewer exploiters than the total number of exploiters in the group. This, in turn, results in more investors at equilibrium than otherwise (Caraco and Giraldeau 1991; Vickery et al. 1991; Hamilton 2002; Afshar and Giraldeau 2014).

Alternatively, with small distances between patches, population density will be higher. This will result in higher encounter rates for investors and hence fewer exploiters (Hamilton 2002), but also higher detection rates for exploiters. In addition, the relative ratio of investors to the number of patches will affect how many investors are present on a patch, in contrast to the basic game where the number of patches is assumed to be considerably larger than group size. Under such conditions, the finder's share will be decreased, and the population's equilibrium will be shifted toward more exploiters than otherwise. Finally, resources might be characterized by both spatial structure and temporal variability in quality. Here, the resulting dynamics are likely to be complex, and how the population's equilibrium would be affected by such characteristics remains to be investigated.

Population Structure

In many models of the PS game, interactions between individuals are assumed to occur at random. Real-life populations are, however, often characterized by some degree of structure, either spatially or in terms of strength of ties between interacting agents. Therefore, interactions do not occur at random. Structured interactions are expected to have a considerable impact on population-level outcomes. For example, Mathot and Giraldeau (2010) showed that in groups of related individuals, the average higher relatedness led to higher proportion of exploiters at the equilibrium, because investors tolerated exploitation from relatives but imposed costs (via aggressiveness) on unrelated scroungers. Likewise, if different groups have different sizes, which can change over time, this can alter the equilibrium frequency of investors. More research is needed, however, to understand fully how dynamic group sizes affect the PS game. Furthermore, it has been shown that group augmentation, whereby groups grow until reaching some maximum size for their environment (carrying capacity), can actually enhance cooperation (Kokko et al. 2001). Therefore, an increased frequency of investors could be predicted in populations that are still growing or have not reached their maximum carrying capacity.

Interactions within groups are also not random because different individuals occupy different positions within groups and social networks. Individuals within the center of groups have reason to be more exploitative, because there are more potential investors for them to exploit, and more potential exploiters to take advantage of their own investment. In contrast, peripheral individuals are more likely to benefit from investing in their own production (Barta et al. 1997). To test these predictions, Flynn and Giraldeau (2001) trained a subset of individuals within groups to be investors in an experimental study of captive ground-feeding nutmeg mannikins (*L. punctulata*). In support of the theory, they found that individuals with the ability to invest were more likely to adopt peripheral locations, whereas exploiters tended to adopt more central positions.

Defining the Individual: When Teams Compete

The interests of individuals and groups, or more specifically, among the individuals within groups, are often in conflict. For example, an individual benefits if others pay more taxes. Conflicts of interest also occur between subgroups of a larger group, as in when business tries to enforce patents over generic pharmaceuticals, or when different nation-states negotiate a treaty to limit anthropogenic climate change. In the tax example, individual interests compete; in the latter examples, business or international interests compete. Therefore, as there are opportunities for exploitation in both, it may at times be appropriate to consider multiple individuals acting together as a single entity within a game played at a higher level. Conversely, such higher-level individuals may be weakened by an internal game between investors and exploiters. It would be useful to know if the dynamics of the PS game apply equally well regardless of the scale of the competing units.

The cognitive abilities of humans to enter into contracts means that groups of individuals may be very effective in their collective strategy (investing or exploiting) within a higher-level game. For example, Milinski et al. (2016) show that individuals within experimental groups that mimic nations involved in negotiating issues related to climate change will select leaders that exploit on their behalf at the international level. Another example is members of street gangs, who like a pride of lions cooperating to hunt large game, may collaborate to rob other individuals. However, even within the gang, there may be greater investors, who put in most effort or perform the most risky part of the jobs, and exploiters who pocket a share of the gang's proceeds without having shared in the risks. This shows that whether an individual should be classified as an investor or exploiter depends on the level of competition upon which we focus. A Mafioso may be seen as an investor on the lower level where he contributes to the activities of the gang, which as a group constitutes a bunch of exploiters at the higher level of society as a whole (see Foster et al. and King et al., both this volume).

The forces that determine equilibrium at one level effect its impact at higher levels. This is because within-level dynamics lead to the group being less effective at whatever it is doing than it could be in the absence of such an internal game. For instance, fishermen on the same boat have to cooperate to maximize their harvest, but one individual may save energy or avoid danger by shirking work. At the individual level, this fisherman is exploiting the efforts of his crewmates. However, if the boat is competing with other boats to harvest rapaciously a shared patch of sea, then any boat that restrains from maximal captures of fish actually invests in the replenishment of fish stocks. In other words, the unmotivated fisherman who was an exploiter at the individual level becomes an investor in the game between boats (see Valone et al., this volume). It is thus important to always be clear about the level of competition under analysis and who is responsible for the decisions in the "game."

Cognition and Learning Behavior

Game theoretical models of social interactions are vastly simplified depictions of animal behavior. The robustness of any results deriving from such simplified depictions needs to be tested by using models that incorporate more mechanistic aspects of behavior (McNamara 2013). For instance, are individuals able to learn over time whether it is better to be an investor or exploiter, or do they have a fixed disposition (Afshar and Giraldeau 2014)? If individuals (or teams) learn, then what information do they use to update their strategies? Do they rely on personal experience (individual learning), or do they learn by observing the behaviors, and perhaps success, of others (social learning)? Here we explore how flexible investor–exploiter tactics are as well as how animals attend to uncertainty, and consider the implications of both individual and social learning in humans and nonhumans.

Individual Consistency

One issue that arises from social foraging experiments and observations is whether investors and exploiters are different types of individuals, or whether these are just different tactical roles that an individual may adopt under different circumstances. The answer appears to be a bit of both. Experiments have shown that individuals appear to shift from being an investor to an exploiter, and vice versa, when the relative payoffs (or risks involved) change accordingly (Koops and Giraldeau 1996; McCormack et al. 2007; Morand-Ferron et al. 2007, 2011). There is also evidence for some individual consistency in strategy use. However, we lack a clear understanding of what makes some individuals be producers and others scroungers.

The corresponding evolutionary question is: Are all individuals in a population genetically endowed with a similar, but flexible, behavioral strategy that can adjust to local circumstances, or do different individuals have different genetically encoded strategies that are maintained by a process of frequency-dependent selection? One way to answer this would be to investigate if there is any genetic basis to these differences in propensity to adopt an investing or exploiting behavior. For example, a genetic basis has been discovered for the alternative mating strategies of male ruffs, *Philomachus pugnax*, whereby three different, genetically encoded, strategies coexist in the population (Lank et al. 1995; Kupper et al. 2016): males can be aggressive "independents," semi-cooperative "satellites," or "faeders" that mimic female appearance. The complex interactions between these strategies have at times been considered analogous to the PS game, whereby some males exploit the efforts of other males to attract and arouse females (Barnard and Behnke 1990), similar to the male frogs mentioned above that avoid the costs of predation by remaining silent and exploiting the calling efforts of other male frogs.

There is some evidence for individual consistency in strategy use, but it is unclear whether individual consistency reflects a stable individual disposition or is an artifact of a flexible behavior that appears stable in a stable social environment. For example, individuals may adopt a suitable tactic considering their relative social rank, with weaker individuals being exploited by dominant individuals. Evidence that an individual's "type" may arise in response to their position in their social world comes from Morand-Ferron et al. (2011), who showed that simply moving individuals to new groups could erase apparent individual consistency in nutmeg mannikins. Furthermore, McCormack et al.'s (2007) work on Mexican jays showed that although many individual jays consistently used one strategy more than the other, many actually used a mixture of strategies, opportunistically choosing to stop searching for food when a subordinate to themselves, from whom they could seize the food, was searching nearby.

Alternatively, it may be true that investors and exploiters belong to different distinct types, but that these types may be better characterized by other, associated, behavioral qualities. For example, Katsnelson et al. (2011) showed that young house sparrows (*Passer domesticus*) were individually consistent when choosing to invest or exploit, but that consistent use of the strategy was predicted by better performance in a prior, foraging-related, learning task. Here, investing may be an associated behavior of superior foragers, because investing is relatively more beneficial for them than it is for inferior foragers. Katsnelson et al.'s (2011) results support, therefore, Arbilly et al.'s (2010) theoretical analysis: over evolutionary time, a tendency for investor behavior may become coupled with sophisticated, but costly, learning behaviors, and likewise simple but cheap learning mechanisms may become coupled with exploitative behaviors.

Much of the literature in psychology and economics implicitly or explicitly claims that players in games have mutually exclusive "types," with some valuing social concerns more than others (Fischbacher and Gaechter 2010). In social psychology and economics, different methods have been developed to assess how the social value orientations of different individuals vary (Grzelak et al. 1988; Liebrand and McClintock 1988; Fischbacher et al. 2001; Charness and Rabin 2002; Murphy et al. 2011). One of the most common methods involves so-called decomposed games, which remove any strategic concerns from social decision making and aim to measure an individual's concern for others (Murphy and Ackermann 2014). Recent evidence, however, suggests that a nonnegligible share of people may embrace seemingly mutually exclusive dispositions. Studies in Mexico with fishermen and in Namibia with pastoralists found that a large share of individuals were both prosocial as well as antisocial toward their fellow villagers, and that contextual environmental factors at the group level explained variation of this behavioral pattern (Prediger et al. 2014; Basurto et al. 2016).

Addressing Uncertainty in Investor–Exploiter Games

Organisms often have to attend to uncertainty in variable environments. Under a range of scenarios, individuals are expected to be sensitive to the average (mean) gain they can achieve from a behavioral action as well as to respond to variance around that average (Stephens 1981). In contrast, the standard biological formulation of the PS game models foraging returns based on a unique value for each model parameter. For example, how often an animal encounters a patch that contains food (the encounter rate) is modeled as λ, which represents the average encounter rate with food patches. However, producers will sometimes be in an environment where they discover food with encounter rates above the average as well as below the average. Thus, for any given searching bout, an animal experiences uncertainty in the exact time it will take to discover a food patch.

Does this type of uncertainty alter investor–exploiter interactions? The nature of the response to uncertainty will depend on whether the negative consequences of a deviation below the average are greater or less than an equal deviation above the average (Stephens 1981). More generally, several stochastic dynamic models demonstrate that stochastic variation, either in patch richness or patch encounter rates, exerts stronger effects on the variance in intake rates for investors compared with exploiters (Caraco and Giraldeau 1991; Barta and Giraldeau 2000; Afshar et al. 2015). These predictions have been confirmed experimentally (Lendvai et al. 2004; Wu and Giraldeau 2005). Therefore, even though both investor and exploiter tactics receive equal payoffs at the equilibrium frequencies, they can differ in the variance of payoffs they experience.

Although uncertainty is likely to have important consequences for the outcomes of investor–exploiter interactions (e.g., stable frequency of tactics, individual differences in tactic use), there are only a handful of empirical studies that have investigated how investors and exploiters respond to uncertainty. This remains an unchartered area in need of exploration (Lendvai et al. 2004; Wu and Giraldeau 2005; Mathot et al. 2009; Afshar et al. 2015).

Learning in Nonhumans

The evidence discussed above suggests that, at least in some species, individuals learn from experience to adjust their use of the investor and exploiter strategies, rather than relying on a fixed strategy, and that populations will reach (or arrive close to) the ESS through learning. This is similar to the idea of Nash's large population of individuals in "mass action," who use their experience and limited knowledge to gravitate toward the equilibrium (Kuhn et al. 1996).

However, as learning presumably entails costs, a strategy that uses learning would have to be superior to a fixed or randomly mixed strategy, otherwise it will not be favored by natural selection.

If individual agents face variable environments, they may do better to employ learning behaviors that update tactics, depending on both individual and/ or social experience. In the scenario depicted by the PS game, the payoffs of investing and exploiting depend on two quantities: the outcome of each strategy when interacting with each possible alternative, and the frequency with which the strategies are present in the population. This raises the question of how the behavior of agents can be tuned to these two categories of information. The solution depends on the nature of the agents being considered.

Simple psychological mechanisms of reinforcement that reward profitable behaviors can lead to an increase in the actions that are, here and now, best. For instance, as rich patches are depleted, or a large number of individuals are driven to exploit rather than invest in the search for food, the experienced payoff by each individual will shift. Thus, if individuals respond appropriately to the changes in their experienced payoffs, the incidence of each kind of action in the population will shift dynamically in the direction that an outside observer with full knowledge might predict.

Several experiments in birds have shown that individual experience affects strategy choice. Nutmeg mannikins have been shown to adjust their use of the exploiter strategy in response to the distribution of food, but previous experience affected how quickly and accurately they adjusted to the new condition (Morand-Ferron and Giraldeau 2010). Perhaps the most direct evidence of strategy-use learning comes from an analysis of strategy choice in house sparrows. Belmaker et al. (2012) found that individuals were more likely to use a previously experienced strategy that had yielded a higher success rate (Belmaker et al. 2012). Similarly, Katsnelson et al. (2008) showed that individual experience of different social environments can affect strategy choice in socially foraging house sparrows. Specifically, hand-reared house sparrows that experienced a "productive mother" (a stuffed female sparrow that frequently made food available by only pecking in places where there was food) were more likely to later adopt an exploitative strategy.

The mathematical description of how animals update memory with their experience of alternative strategies (learning rule), and how they choose between strategies based on these memories (decision rule), has been debated over recent decades. A number of models have tried to identify the evolutionarily stable learning rule: the mathematical rule which, much like the ESS, once fixed in the population, cannot be invaded and replaced by any other rule (Harley 1981; Tracy and Seaman 1995). By implication, this rule should allow learners to reach the ESS or at least approximate it. More recent work has used agent-based simulations to find the evolutionarily stable learning rule in an investor–exploiter framework (Beauchamp 2000; Hamblin and Giraldeau 2009). In these models, learners continuously modified their strategy based on previous experience, and their strategy choice at each time step affected the experience of others in the population. These studies identified some learning rules as more evolutionarily stable than others. Still, since the poor performance of some of these rules can be resolved through coupling with flexible decision rules (Arbilly 2015), the evolutionarily stable learning rule in the PS game has yet to be determined. Since there are likely various learning processes that may converge to the ESS (Selten and Hammerstein 1984), fitting into different learning models detailed data of the behavioral choices animals make following experience might be the best way to identify these rules.

While learning may seem like the best way to approach the problem of a changing ESS, evolutionary models of the PS game that compare strategy learning with innately fixed strategies suggest that the advantage of learning is not straightforward. The social, rapidly changing environment presents a serious challenge: when everyone adjusts their behavior based on previous experience, previous experience may become irrelevant. Explicit modeling of the learning process in the PS game, for example, revealed that learning can be favored only in a fast-changing (physical) environment or when individuals have some preexisting trait that makes them perform better in one of the two strategies (Katsnelson et al. 2011). Furthermore, since learning is presumably costly, it may be disfavored once the population reaches a stable equilibrium, because learning is no longer needed (Dubois et al. 2010).

Learning in Economics

In many of the scenarios under discussion, individuals interact repeatedly for some indefinite amount of time. Such repeated interaction gives a potentially important role to cognition, as human actors can theoretically anticipate the future behavior of their "opponent." By reasoning what their opponent will do in the final round of a series of interactions, individuals can work backward to choose their current actions accordingly, using "backward induction" (Von Neumann and Morgenstern 1953; Aumann 1995; Binmore 1996). For example, imagine a scenario where there is only one round of interactions between two players, and that not cooperating is individually the most attractive option even though mutual cooperation is more attractive than mutual exploitation (the prisoner's dilemma) (Rapoport and Chammah 1965; Axelrod and Hamilton 1981; Tucker 1983). Here, the Nash equilibrium, the best response by any one player to the actions of the other, is to not cooperate (to exploit the other). If this scenario is repeated indefinitely, it can pay to invest in cooperating as the benefits of a long-run cooperative relationship can outweigh the short-term gains of exploitation, and there can be many Nash equilibria in repeated games (Kreps et al. 1982). However, if the number of interactions is common knowledge, then individuals can use backward induction to reason that their partner will exploit them in the final round. Thus, they may as well exploit them in the preceding round rather than invest in cooperating, and so on all the way back to the opening interaction (although see Pettit and Sugden 1989).

For another example of backward induction undermining cooperative strategies in repeated games, consider the strategy of grim trigger (Axelrod 2000). This strategy starts out by choosing cooperative actions, such as not using nuclear missiles, but threatens to switch permanently to noncooperative actions after the opponent has been observed using exploitation (e.g., using nuclear missiles). The threat of this strategy serves to deter exploitative strategies and implies that mutual cooperation can be maintained. However, if there is a final round to the game, perhaps due to nuclear annihilation, then there is either no opportunity to retaliate or no incentive to follow through on a costly deterrent. This means the deterrent is no longer a credible threat, and thus exploitative strategies should be used in the final round of the game. Consequently, there is no reward for cooperating in earlier rounds of the game, meaning cooperation is disfavored.

Empirical evidence on whether or not humans use backward induction is mixed (Binmore et al. 2002). Typical evidence from the prisoner's dilemma suggests that backward-induction reasoning kicks in only as the end of an interaction becomes near (Andreoni and Miller 1993; Embrey et al. 2016). This pattern has been explained by learning models that view human actors as adaptive but capable of displaying some degree of foresight (Jehiel 2001; Heller 2015).

Whether or not nonhuman animals should be viewed as consistent with backward induction will largely depend on whether evolutionary selective pressures select for outcomes consistent with backward induction (Noldeke and Samuelson 1993). Learning and cultural transmission also play an important role for humans in acquiring strategies, consistent with approaches such as backward induction (see Mengel and van der Weele, this volume).

Models used to describe learning in economics can require little in terms of cognitive resources from agents. Take, for example, reinforcement or stimulus-response learning, where actions that have led to good outcomes in the past are more likely to be repeated in the future. Agents have a probability distribution over possible actions. When an action is chosen, the probability of that action being taken again rises in proportion to the realized payoff. The action has been "reinforced." Note the very low level of information or processing ability necessary to implement such an algorithm. In the context of game playing, an agent does not need to know the structure of the game to calculate best responses or even to know that a game is being played (Foster and Young 2006; Pradelski and Young 2012; Nax et al. 2016).

More sophisticated models are based on best response behavior and imitation or may involve Bayesian rationality and forward-looking behavior (Fudenberg and Levine 1998). Which of these learning models describes human behavior best under a specific condition is as yet unresolved (Cheung and Friedman 1997; Camerer and Ho 1999). While this issue has mostly been addressed within the context of a single game, some authors have recently started to acknowledge the importance of understanding learning beyond the context of a single game (Jehiel 2005; Mengel 2012).

When trying to understand how these models extend to nonhuman actors, we would like to know how well the outcomes of these models could be described as outcomes of evolutionary processes. In the case of some of these learning models, such as reinforcement learning, it is well known that close links to the evolutionary replicator dynamics exist (Taylor and Jonker 1978; Börgers and Sarin 1997; Hopkins 2002). Hence, not only can the outcomes of these models of differing cognitive demands be well described by evolutionary models, we should also expect Nash equilibria to be played. However, as some humans arguably use more complex learning rules than nonhumans, it is less clear whether we should expect nonhuman animals to reach the same outcomes as humans (cf. Lange and Dukas 2009).

Innovation and Technology

Individual learning is required for innovation but can be costly, requiring investments of time, energy, and perhaps physical materials. A fundamental problem for the innovator is thus whether these investments can be recouped through the advantages that stem from the innovation. Consequently, there is a strong incentive for other agents to learn socially and to copy the innovation of others, rather than invest in innovations. When the benefits of an innovative behavior stem from increasing relative competitive ability, copying of such behavior can quickly erode any advantage, decreasing the incentives for innovation in the first place. While social learning can reduce the incentives for innovation and the number of potential innovators, it can also facilitate the transmission of innovative behaviors through social networks. Thus populations with a blend of individual and social learning may exhibit a high level of cultural developments, despite the negative aspects of copying (Rogers 1988; Boyd and Richerson 1995; Kendal et al. 2009; Rendell et al. 2010).

The temptation to copy is central to the economics of innovation. Societies aiming to promote innovation have therefore mitigated the costs of being copied by granting temporary monopolies in exchange for the dissemination of innovative technologies. The most prominent economic institutions are patents and copyrights; however, these institutions can have the undesired effect of granting the innovator a monopoly position that can be used to charge high prices, temporarily reducing the availability of new technologies. The trade-off between the incentives for innovation and the ensuing market distortions has been the topic of a large literature, and many different solutions are proposed and observed in practice. For example, the Creative Commons is a nonprofit organization that simplifies the process of copyright and reduces the costs for copyright owners and licensees, facilitating the sharing of useful intellectual property.

Models investigating the potential evolution of innovation using the PS game, where investors are innovators and exploiters are copying their innovations, reaffirm the essential role of patents and copyrights in sustaining innovative behavior. If we assume that innovations may be advantageous as well as disadvantageous, the presence of innovators in the population may greatly fluctuate; their innovations can lead to a substantial increase in population size, but can also result in the extinction of the population altogether (Lehmann and Feldman 2009). Computer simulations of evolution suggest that innovators are not likely to persist for more than a few generations in a population of copiers, unless there is some penalty to copying; for instance, when behavior is not replicated faithfully or some social reward (e.g., prestige or royalties) is given to innovators. Altogether these models suggest that the ability to enforce social reward through institutions may be fundamental to the high rate of innovations in humans compared to other animals.

Technology may serve different purposes for investors and exploiters. Investors may use technology to obtain resources or to protect themselves from exploiters, whereas exploiters may employ technology to track the behavior of investors. For example, some fishermen may invest in technology for locating shoals of fish while others may invest in technology to locate those fishermen's boats and exploit whatever they find. This situation may become unstable. Evolutionary simulations have shown that if the competition between strategies leads to escalation, where each strategy is investing increasingly more in technology to outwit the other, the emerging arms race proves, in the long run, to be unstable. It is likely to result in either the extinction of exploiters, if the investors manage to race ahead and open a large enough gap, or in a slow backward arms race if the exploiters race too far ahead so that investors are better off investing less in their technology (Arbilly et al. 2014).

Conclusion

The benefits of investment into the production of a "good" are undermined by the risk of others exploiting such investments. This risk creates a dilemma for potential investors in a host of real-life situations that are faced by many animal species, including humans. In this chapter, we have demonstrated that the essence of this dilemma is well captured by the so-called PS game, a simple but powerful biological framework for investigating situations of conflicts of interests between individuals. We have identified numerous biological examples, such as foraging behavior in birds and primates, as well as examples in various human societies: from small-scale communities managing resources, to firms investing in research and development, and nations competing over endangered natural resources. Crucially, in all these examples, we have highlighted how the proportion of resources, from which investors can benefit before exploiters deplete the resource (i.e., the finder's share), is key to the population level of exploitation. In both animals and humans, interested parties have developed elaborate mechanisms that increase the value of the finder's share (e.g., patents and copyrights in human societies, or resource defense in animals).

Finding cooperative solutions to the type of dilemma captured by the PS game has attracted much attention from both the fields of biology and economics. Although the two approaches often tackle this problem from different perspectives, with specific empirical and theoretical tools, we find that they share commonalities. In both approaches, for example, individuals are modeled as agents seeking to maximize a certain quantity (e.g., biological fitness, economic payoffs, or general utility). Consequently, groups of rational agents cannot often be modeled in the same way as individuals, since the interests of agents within a group are not necessarily aligned with those of the group. In addition, we find that the PS game could be encapsulated within a more general mathematical framework, and have provided a single model to unify the well-known games from economics, specifically the public goods game, the volunteer's dilemma, and the rent-seeking model. Our general framework can be used to extend the range of dilemmas that can be studied with the PS game, for example, in situations where a certain number of investors is required to produce a good or to attract exploiters.

In an attempt to highlight current limitations of the current PS framework, we demonstrated how Elinor Ostrom's well-known classification of variables in the social ecological framework can help identify important aspects of the game theoretic framework that remain unexplored (Ostrom 2007). For example, the value of the finder's share determines the degree of excludability, and thus the typical classification of goods in economics needs to reflect this continuum and be less discreet. In general, characteristics of the resource of interest (rivalry, excludability), the type of population structure of the agents, as well as the different types of agents at play (individuals, teams, firms, or nations) are all likely to have important implications for the dynamics of exploitation within populations. Thus, incorporating these features will extend the value of the PS framework to a wider range of biological and economic scenarios.

The PS game is affected by both the resource characteristics and the nature of the agents involved; however, the implications of certain key factors of the agents remain to be resolved. Whether the agents, both human and nonhuman, are capable of learning and how they attend to uncertainty in more realistic biological scenarios is still unclear but important, because learning can potentially alter the predicted equilibrium between investors and exploiters. The role of learning is particularly important in humans because of our unusual ability for cumulative cultural development, innovation, and sophisticated technology, which allows us to find, develop, and keep novel solutions to the problem of investment versus exploitation. These features, however, also allow potential exploiters to increase their ability to benefit from the investments of others.

Exploitation in the Context
of Natural Resources

4

Producer–Scrounger Models and Aspects of Natural Resource Use

Zoltán Barta

Abstract

Humans are currently using natural resources at unprecedented rates and it is not difficult to extrapolate how this could lead to global catastrophes of various kinds. To mitigate eventual consequences, our understanding of the processes involved must be improved. Since resource use frequently involves groups, free-riding behavior (i.e., exploiting the efforts of others) must be expected. Recent evolutionary studies indicate that exploitation of others' efforts can dramatically alter how resources are utilized. Two types of effort are exploitable: the harvesting and maintenance of resources. This chapter argues that the exploitation of harvesting efforts can be analyzed as a producer–scrounger evolutionary game. The presence of scroungers (exploiters) in a group usually decreases the overall use of resources by the group. Factors that increase the proportion of scroungers (e.g., energy reserves, existence of dominance hierarchy, or prevalence of relatedness) can further decrease resource use. By contrast, aggression and the compatibility of scrounger and producer strategies elevate resource use. In temporally unstable patches, scrounging does not affect resource use in groups that are at equilibrium. Encouraging scrounging may lower resource use, even in humans, but this raises a moral dilemma: individual scrounging is bad, reduced resource overuse by the population is good. Surprisingly, only a small portion of the literature has considered the consequences of cheating in terms of the natural resource management—a situation that demands attention in future research.

Introduction

One of the greatest problems of the Anthropocene (Lewis and Maslin 2015) is the overuse of natural resources, including fisheries, forests, clean water, air, and fertile soil. This not only inhibits humankind from developing strategies for sustainable resource use, it can easily result in ecological, economical, and social catastrophes on a global scale. Nevertheless, decisions are usually made on a local scale, mainly at the level of individuals. To prevent overuse, it is therefore of paramount importance to understand the processes involved in individual resource use decisions.

Most of the natural resources that are threatened by overuse can be classified as common-pool resource systems (Janssen et al. 2010). Similar to public goods, it is impossible or unfeasible to exclude anyone from using common-pool resources (Ostrom 1990). In contrast to public goods, however, resource use by one individual in a common-pool resource system decreases the amount of resources available for others (Ostrom 1990). This situation closely resembles one in which animals forage on patchily distributed food, a situation that has been widely studied in ecology and evolutionary biology. Thus, knowledge on individual decisions accumulated in these fields might potentially provide useful insights to help prevent resource overuse by humans. To facilitate this information exchange, I review the recent theoretical advancement in a special case: group foraging.

Studying group foraging is important for the following reasons: First, in many cases, both in humans and other organisms, resource use takes place in groups. In humans, for instance, local communities utilize the surrounding forests or fleets of ships exploit fish stocks. Many animals have also been frequently observed to forage in flocks (birds), schools (fish), or herds (ungulates). Second, models of group foraging consider individual decisions and, through game theory, explicitly take into account the interactions between individuals. Interacting with others can affect the dynamics of resource use by altering its costs and benefits. For instance, the benefits of investing in a local timber industry can be greatly reduced when the harvested timber is stolen. An animal analogue can be found in ground-feeding passerine birds, like sparrows, where some individuals invest in actively searching for new food patches, while others simply wait for a "neighbor" to find a patch so that they can rush in to obtain a share of the food, obviously decreasing the food intake of the patch finder (Giraldeau and Caraco 2000). At the individual level, against a backdrop of possible exploitation, the cost of investors (producers) and the benefit of exploiters (scroungers) can result in a fewer number of individuals willing to invest in the production of resources (timbers, food patches); this, in turn, may reduce the level of resource use by the group as a whole. Therefore, if the group is overusing its environment, then the spread of scroungers can mitigate the problem of overuse. The processes underlying this scenario are well captured, at least in evolutionary biology, by the theoretical framework of producer–scrounger (PS) games (Giraldeau and Caraco 2000). In this chapter I present a baseline model for producing (investing) and scrounging (exploiting) and investigate how the presence of scroungers influences a group's intake of resources. Thereafter I review models that are extensions of the baseline model and explore how these modifications alter resource use. I conclude with a brief investigation of how changing resource characteristics can influence the effect of scroungers on resource use by the group.

The Baseline Producer–Scrounger Game

Following Giraldeau and Caraco (2000), let us consider a group of G individuals foraging for T time units in an environment where food occurs in well-defined patches. A food patch contains F food items. We assume that the time needed to consume a patch is negligible compared to the time needed to find the patch because patches are difficult to locate. To obtain food, individuals must either invest in searching for patches or exploit the search effort of others. This is modeled by assuming that individuals can follow either a *producer* or a *scrounger* tactic. An individual playing the producer tactic actively searches for food and finds patches with rate λ. After finding a patch it consumes a portion a of the patch while alone (the finder's advantage, $a \leq F$) and shares the remaining food ($A = F - a$) with the arriving scroungers. The proportion of individuals adopting the producer tactic (producers) in the group is p. We assume that a producer never feeds from patches found by other producers, whereas scroungers only feed from patches found by producers. In other words, the tactics are incompatible (Coolen et al. 2001). Finally, we assume that scroungers are able to detect all patches found by producers and thus all scroungers can feed from each patch found.

Under these assumptions, the food intake for a producer in a group containing $(1 - p)G$ scroungers is:

$$W_p(p) = T\lambda\left(a + \frac{A}{(1-p)G+1}\right), \tag{4.1}$$

whereas a scrounger's intake is:

$$W_S(p) = pGT\lambda\frac{A}{(1-p)G+1}. \tag{4.2}$$

At evolutionary stability, $W_p(\hat{p}) = W_S(\hat{p})$ and hence the evolutionarily stable frequency of producers, $\hat{p}$, is:

$$\hat{p} = \frac{a}{F} + \frac{1}{G}. \tag{4.3}$$

This means that the proportion of producers increases with the proportion of food available exclusively for them (the finder's share, a/F), and decreases with group size G. Accordingly, the two strategies can coexist: $\hat{p} < 1$, if $a/F < 1 - 1/G$. The equilibrium point is stable because of the strong negative frequency dependence of the scrounger's food intake (for details, see Giraldeau and Caraco 2000).

The average per capita intake in a group of G producers foraging independently (see above) is λTF amount of resources. Thus the amount of resources used by an average individual in a group of pG producers and $(1 - p)$ G scroungers is:

$$pW_p(p) + (1 - p)W_S(p),\qquad(4.4)$$

which simplifies to $p\lambda TF$. Accordingly, the use of resources, not surprisingly, decreases as the proportion of producers decreases, and hence the proportion of scroungers increases in the group (Vickery et al. 1991; Giraldeau and Dubois 2008).

Let us now investigate what happens in groups at evolutionary stability. To obtain the per capita intake in a group, where the proportion of producers is at the evolutionarily stable value $\hat{p}$, we substitute p with $\hat{p}$:

$$\left(\frac{a}{F} + \frac{1}{G}\right)\lambda TF.\qquad(4.5)$$

From this it follows that the evolutionarily expected per capita resource use in equilibrium groups decreases when group size increases and finder's share decreases. This is not surprising because these are exactly the conditions that facilitate an increased number of scroungers at evolutionary stability (Figure 4.1).

According to the reasoning above, if a population is partitioned into multiple small foraging groups, the total rate of resource harvesting will be higher than if individuals had foraged in just a few but larger groups. Furthermore, if

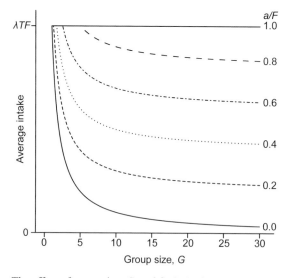

Figure 4.1   The effect of group size, G, and finder's share, a/F, on an average individual's intake in groups containing an evolutionary equilibrium proportion of scroungers. λTF marks the average intake in a group of pure producers.

the proportion of a patch available exclusively to its producer, a/F, is small, then the overall rate of resource harvesting decreases. The proportion of a/F can be small if patches are large, because relatively small amounts will be consumed by the producer before the arrival of scroungers. A high density of foragers can also result in a small a/F because high densities mean that individuals are close to each other, and hence scroungers can quickly reach a discovered food patch, leaving just a short time for its producer to consume the patch alone (Giraldeau and Caraco 2000).

By combining reasoning similar to that presented above with population dynamics, Coolen et al. (2007) pointed out that the spread of scroungers decreases the use of resources. Furthermore, they found that the coexistence of scroungers and producers among predators leads to stable population dynamics in a simple prey–predator system instead of the more usual cycling. As a consequence, they argue that prey species should be expected to evolve in such a way that facilitates scrounging in its predators. One of the possibilities they envision is that prey become more cryptic, as this increases the cost of searching and makes scrounging a more appealing option. The other possibility is that prey occur in large patches, which results in an overall smaller finder's share; this, in turn, will increase the benefits from scrounging (Coolen et al. 2007).

Extensions of the Producer–Scrounger Game

This baseline PS model cannot, of course, address all of the complexities inherent in the social foraging process. For instance, because all individuals are treated the same, one cannot know how differences in energetic state, dominance rank, or the possibility of aggressive resource defense would influence the spread of scroungers. Taking individual differences into account, however, is important, because this makes it possible to predict the characteristics of both producers and scroungers. This knowledge, in turn, might allow the manipulation of resource users, and hence the volume of resource use.

The economic state of agents can vary widely and influence their decisions. For instance, a person close to bankruptcy values items differently than someone who is well off financially. Animals, too, can differ in their state, one of the most important differences being that of energetic state. Differences in energy levels (energetic state) can simply arise because of the inherently stochastic nature of the foraging process: individuals usually collect different amounts of energy. The energetic state can be an important determinant of behavior for two reasons (Houston and McNamara 1999):

1. It can constrain the available behavioral options for an individual: an individual with a low level of energy reserves cannot afford to rest in safety from predators, it must forage to avoid starvation.

2. It can influence the value of food: a given amount of food is worth less to an animal with a high level of energy reserves than to one close to starvation.

Thus, it is expected that energy reserves affect the costs and benefits of the social foraging process, and hence the use of the producer and scrounger tactics. To investigate this effect, Barta and Giraldeau (2000) developed a state-dependent dynamic PS game. With this model, they considered a group of foragers that needed to survive several winter days and nights. They found that the use of the scrounger tactic depends on both the levels of energy reserves and the time of day. Early in the morning, individuals with a low energy level utilized the scrounger tactic whereas those with higher reserves used the producer strategy. Later in the day, however, this pattern reverses: individuals with high reserves tend to scrounge while those with low reserves produce. The reason for this pattern is rooted in the variance-sensitive properties of the tactics: the producer tactic is a variance-prone tactic so individuals using this tactic can get high amounts of food (as long as $a > 0$); however, they rarely achieve this level rarely because producers can only feed from patches they find themselves. On the other hand, use of the scrounger tactic can be a variance-averse alternative: scroungers get small amounts of food (because they have to share the patch), but frequently (as long as there is more than one producer in the group). Early in the day many individuals tend to have low energy reserves, because they have just survived the long winter night; to avoid starvation they need a small, but reliable amount of food. Later in the day animals need high reserves to survive the night. Those who are close to this limit play it safe—they use the scrounger tactic—while those who have low reserves late in the day must take risks and use the producer tactic. This policy results in a high frequency of scroungers early and late in the day, while during the middle of the day the proportion of scroungers is lower. Importantly, this state-dependent dynamic PS game model indicates that the availability of the scrounger tactic is not necessarily a cost of group foraging, as was assumed previously (Vickery et al. 1991), but may be advantageous as it can provide insurance against starvation in a stochastic world. Because of this insurance characteristic, the use of the scrounger tactic is more advantageous in this game than it is in the baseline PS game. As a result, the proportion of scroungers at evolutionary equilibrium is expected to be higher in the state-dependent game than in the baseline game, especially under medium values of finder's share. This means that average harvest rates decrease in a state-dependent world, while the survival of individuals increases when the scrounger tactic is available (Barta and Giraldeau 2000).

Dominance relationships and the dominance hierarchy that emerges are inherent parts of group life. Some individuals in a group commonly gain larger shares of limited resources (e.g., food, safety, or mates) than others do. For example, a fishing vessel fitted with a more powerful engine can pull larger fishing nets and hence get a larger share of a fish school than a vessel that is not so well equipped. In animals, based on this asymmetry in competitive ability, individuals in a group can be ordered in a dominance hierarchy, where dominant individuals have stronger competitive abilities and hence dominate those, the subordinates, who have weaker competitive abilities. A consequence of this competitive asymmetry is that the dominance status of an individual should have a considerable effect on the individual's use of social strategy. Indeed, in Harris's sparrows (*Zonotrichia querula*), dominant individuals frequently supplant subordinates from food patches that have just been found (Rohwer and Ewald 1981). To investigate systematically how the magnitude of differences in competitive ability between individuals in a group influences social foraging decisions, Barta and Giraldeau (1998) developed a phenotype-limited model of producing and scrounging. In phenotype-limited games, individuals differ in some respect and this influences their gain from the use of different strategies. In the model by Barta and Giraldeau (2000), individuals differ in their competitive weights, which determine their share from a divided food patch. If the level of difference between the competitive weights of individuals is high, then group members vary considerably in their competitive ability; high-ranking, dominant individuals get a disproportionately large share from a divided patch. If, on the other hand, there is a low level of difference, group members are more or less the same and thus they receive about the same amount from a shared resource. Barta and Giraldeau found that when group members are more or less the same, producing and scrounging tactics are used in the same way; that is, individuals with different dominance rank do not differ in terms of the proportion of scrounging. Increasing the level of difference leads to a region of the parameter space where the relationship between dominance rank and use of scrounging is rather variable. If the level of differences in competitive ability is high, then dominant individuals use scrounging while subordinate individuals use producing exclusively (see also Hamilton 2002). When this strong correlation between dominance rank and tactic use exists, the proportion of scroungers in the group drops remarkably. In turn, this results in a higher than average food intake compared to groups where phenotypic differences do not influence tactic use (i.e., where the level of differences between individuals is low).

An issue inherently related to competition is aggression. The models considered above assume peaceful scramble competition; that is, resources are divided among individuals either equally or proportionally to their competitive weight, but without costly fights. Real animals, however, frequently defend resources aggressively. Aggression, in forms of chemical or viral warfare, is also often observed in bacteria (Brown et al. 2009a). Since aggression can significantly alter the costs and benefits of different social foraging tactics, it is important to investigate how the possibility of aggressive behavior influences the use of producing and scrounging. By embedding a hawk–dove game (a variant of the snowdrift game) into the baseline PS game, Dubois and Giraldeau (2005) presented a model where individuals can decide not just about whether to produce or to scrounge but whether to defend the resources aggressively or not. Escalated fights are costly in terms of energy, time, and elevated predation risk. Because producers have exclusive access to part of the discovered patch (the finder's advantage), while scroungers do not, the authors assumed a role asymmetry between producers and scroungers regarding the aggressive defense of resources. In addition to this role asymmetry, producers and scroungers were assumed to have the same fighting ability; that is, the probability to win the fight and obtain resources was the same for all foragers. Using an iterative method to solve the games, Dubois and Giraldeau (2005) found that producers always defend the discovered patch aggressively, whereas the level of aggression by scroungers depends on the circumstances (see also Dubois et al., this volume). According to Dubois and Giraldeau (2005), the level of aggressiveness differs between tactics: because producers gain more from a patch, they can afford to mount a more intense defense. This makes scrounging a less valuable option. As a consequence, the proportion of scrounging decreased and hence the use of resources increased in this game, compared to the baseline PS game. This analysis, however, did not take into account that the value of the patch might differ for producers and scroungers. One could argue that producers value the part of the patch that is going to be shared with the arriving scroungers less than the scroungers, because producers have already consumed the other part of the patch (McNamara and Houston 1989). Therefore, it might not be entirely unreasonable to assume that scroungers might behave more aggressively to obtain a share from the patch. This, of course, would change the prediction of this model. To settle this issue, a state-dependent analysis of the problem should be conducted.

A crucial assumption of the baseline PS model is the complete incompatibility of producer and scrounger tactics. In other words, a producer cannot recognize the food findings of other producers and a scrounger cannot find a patch alone. Imagine a fishing vessel that is equipped both with sonar (useful to locate schools of fish) and radar (to locate other ships) equipment. The incompatibility of producing and scrounging corresponds to the case when our imaginary fishing boat has a weak engine so it cannot power both the sonar and the radar at the same time; thus, the crew must decide to power the sonar (and hence play producer) or power the radar (and scrounge). Vickery et al. (1991) relaxed this strict assumption by introducing a third tactic, the *opportunist*, into the baseline PS game. To become an opportunist, the owner of the above fishing boat must invest in an engine that is strong enough to power both the sonar and the radar equipment at the same time; hence the boat can simultaneously look both for fish and other fishing vessels. According to Vickery et al. (1991), an opportunist can find food itself. However, its efficiency at locating patch c might be smaller than the food-finding efficiency of a pure producer: $c \leq 1$. As we have seen above, a producer can find patches with rate λ. With this formulation, the rate of patch finding by an opportunist is $c\lambda$. An opportunist is also able to detect scrounging opportunities with efficiency h, which can, again, be smaller than the efficiency of a pure scrounger: that is, $h \leq 1$. This means that a scrounger can detect all other individuals who have found a food patch, whereas an opportunist can only detect a proportion of h of those food-finding events. According to this notation, a pure producer can be characterized as an opportunist, with $c = 1$ and $h = 0$, whereas a pure scrounger would be depicted as $c = 0$ and $h = 1$. This is the case of complete incompatibility. The case when both $c = 1$ and $h = 1$ constitutes complete compatibility; that is, a forager can freely switch between producing and scrounging without any loss of efficiency. This might work in an ideal world, but in reality some cost of switching is expected. As a consequence, c and h should correlate negatively; any increase in one of the efficiencies should result in a decrease in the other (Vickery et al. 1991). This is what Vickery et al. refer to as partial compatibility, of which they distinguish three types (see Figure 4.2a):

1. Exact compensation, when an increase in one efficiency results in the same level of decrease in the other efficiency: $c + h = 1$
2. Overcompensation, when gain in one efficiency leads to less loss in the other: $c + h > 1$
3. Undercompensation, when gains are smaller than losses: $c + h < 1$

Vickery et al. (1991) found that opportunists can only spread under the condition of overcompensation. When undercompensation occurs, just the producer-only and the producer–scrounger combination is stable, depending on the finder's share, as in the baseline PS game (Figure 4.2b). Under overcompensation, four regions of coexistence can be identified along the level of finder's share. At the lowest values of finder's share, opportunists coexist with scroungers. At immediate values of finder's share, pure opportunist is the evolutionarily stable strategy. At higher finder's share, opportunists coexist with producers. At highest finder's share values, producers dominate. With increasing level of overcompensation, the region of pure opportunists expands while the others shrink. The average intake increases with finder's share in the mixed regions but remains constant in the pure regions (Figure 4.2b, c). Increasing the level of overcompensation results in an increased average intake, but opportunists only reach the intake level of pure producers under complete compatibility ($c + h = 2$). Conditions which favor overcompensation decrease the cost of producing as well as the benefit of scrounging, and hence lead to a higher level of resource use by the whole group. Nevertheless, the above analysis does not consider the costs of making overcompensation itself possible. These costs might be substantial, as our analogy of fitting a more powerful engine (to enable the simultaneous use of sonar and radar equipment) into the fishing boat suggests. How these costs affect the equilibrium of strategies and level of resource use requires further investigation.

The incompatibility of tactics seems to be well supported in birds, where the different head positions required for searching for food (head down) and

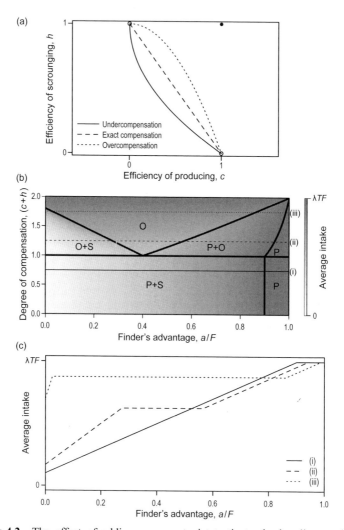

Figure 4.2 The effect of adding an opportunist tactic to the baseline producer–scrounger game on the average intake. (a) Three types of relations are depicted between the opportunist's efficiency of producing a patch, c, and scrounging a patch, h. Open circles represent the complete incompatibility of producer and scrounger tactics; the solid circle shows complete compatibility. (b) The average intake of a forager in equilibrium groups is shown as a function of a finder's share and the degree of compensation $(c + h)$. Shading indicates intake: the darker the shading, the higher the intake. Undercompensation occurs if $(c + h) < 1$, exact compensation if $(c + h) = 1$, and overcompensation if $(c + h) > 1$. Black thick solid lines separate regions of groups of different evolutionarily stable strategy composition. Letters indicate the strategic composition of the group at equilibrium: P, producer; S, scrounger; and O, opportunist. Thin black (solid and dashed) lines marked by (i), (ii), and (iii) refer to the average intake as plotted in (c). (c) The average intake is shown as the function of finder's share at different levels of compensation, marked by the thin black (solid and dashed) lines in (b).

Of course we would not be human if we did not consider our own species when observing such behavior in the natural world, and to wonder how people respond to the decision to invest or to exploit. The problem of investment versus exploitation has also been studied extensively in the various social sciences, such as economics. Not only does the tension between investing and exploiting affect individuals, it can also be seen as a decision facing rival companies and even rival countries. For example, should a firm invest time and money on developing new products, such as a technologically advanced mobile phone, or perhaps a new pharmaceutical product, when it could simply wait for others to do the innovating and then copy them? Should governments order their ships to refrain from overfishing, to allow the replenishment of nearby stocks, at the risk of seeing the fish harvested by ships from another country? When comparing these scenarios to biological examples, one sees a key difference: such decisions play out on a different stage to the ecological stage of biology, and thus are affected by the decisions and institutions of governments and other interested parties. It may be that many of our laws and modern sociopolitical institutions have been shaped by a recognition of the fears of would-be investors (e.g., a permanent, paid police force to protect our bodies, families, and homes, or patent offices to protect our intellectual "property").

In this chapter, we present a general overview of how the problem of investment versus exploitation is conceived, analyzed, and empirically tested, by both biologists and economists. We also consider how the problem can be advanced and what we still need to know. The following discussion investigates if and how the concept of investment and exploitation can be applied to real-world problems. But first, we need to clarify what kind of exploitation we are focusing on here.

If interested in discussing exploitation, one could be concerned with instances where one party exploits another in the sense that they take advantage of another's misfortune, desperate situation, or current weakness. For example, a payday loan company that lends moderately small amounts of money in exchange for very high fees could be considered to exploit desperate people that arguably have no choice. The morality of such behavior is far from clear, as reflected by the mix of laws that try simultaneously to allow but limit the severity of such practice. However, here we are not analyzing the morality of usury, or even investment and exploitation. Instead, we are interested in what affects the "choice" to be an investor or an exploiter, and our analytical approach requires us to be able to clarify different situations.

There are at least two key distinctions that allow us to clarify the difference between the payday loan example and those examples that interest us. First, the customer, legally speaking at least, always has the choice to accept the deal on offer or to walk away. Such interactions are thus perhaps best viewed as a form of negotiation during bargaining, with successful agreements only applying to outcomes that are perceived as beneficial by both parties. However, the reality searching for food-finders (head up) rule out the simultaneous use of such tactics (Coolen and Giraldeau 2003).

Social behavior can strongly depend on the level of within-group relatedness (Hamilton 1964), a feature that is also not incorporated into the baseline PS game. If the group consists of relatives, it is expected that the level of exploitation of efforts decreases compared to a non-kin group (Frank 2003). Others, however, argue that the effect of relatedness should depend on the costs and benefits of scrounging (Tóth et al. 2009). If scrounging imposes a high cost on producers, then high relatedness should decrease the proportion of scroungers. Alternatively, if individuals gain a substantial benefit by scrounging, then scrounging should be more common in groups of relatives than in groups of strangers. Mathot and Giraldeau (2010) modeled this proposition formally. They assumed that inclusive intake rate is a surrogate of inclusive fitness and modified the baseline PS game to include an indirect benefit for the producers and an indirect cost for the scroungers. The indirect benefit is the amount of food gained by the scroungers from the producer discounted by the within-group relatedness. The indirect cost takes into account that scroungers, by consuming part of the patches found by their relatives, decrease their relatives' intake. In the baseline PS game, producers have no options to control the level of scrounging in the group. To overcome this deficiency Mathot and Giraldeau (2010) assumed that producers can impose a cost on scroungers (e.g., through aggression). By increasing the magnitude of this cost, they were able to increase the level of control the producers exert over joining a discovered food patch by others. The authors found, not surprisingly, that increasing the cost imposed by producers on scroungers increased the equilibrium proportion of producers. For a given level of cost, increasing the level of relatedness within the group resulted in higher equilibrium proportion of producers, which, in turn, led to more intense resource use. Interestingly, if producers can discriminate between kin and non-kin scroungers by imposing higher cost on non-kin, the equilibrium proportion of producers can be lower in groups composed of kin than non-kin (Mathot and Giraldeau 2010). In other words, kin groups might contain a higher proportion of exploitative individuals than non-kin groups. The investigation of the role of social preferences (Charness and Rabin 2002) in producing-scrounging decisions might be an interesting extension of this modeling framework.

As discussed above, the spread of scroungers decreases resource use by the group. Nevertheless, all models reviewed so far have been based on the assumption that adding more individuals to the group decreases the intake of group members. Now let us relax this assumption and investigate how the spread of scroungers affects resource use by the group under the following condition: adding a new individual to the group does not decrease the intake of group members. This situation arises when animals forage on ephemeral patches; that is, when food patches disappear before all food items are consumed by the foragers (Barta and Giraldeau 2001). Swarms of flying insects or schools of fish are good examples of this type of food resource. The fashion industry might provide a human example for ephemeral resources (Giraldeau, pers. comm.). In the world of fashion, a design is valuable for only a limited period of time because new designs appear annually. If the producer of a new design cannot supply the market with enough goods due to constraints (e.g., insufficient production or transport capacity), the copiers (scroungers) of the new design will prosper without considerably harming the producer and each other. Resource ephemerality basically transforms the common-pool resource system into a public goods system. To investigate how scrounging influences resource use in such an environment, I present a simple model under the group-foraging scenario used above.

Since producers, by definition, arrive earlier at the patch, they can consume more food than scroungers before the patch disappears. We assume that scroungers have enough time to consume an amount A of food, while producers are able to eat $a + A$ amount. A patch contains F food items. A crucial assumption for the following argument is that patches are not depleted; that is, the food consumed by the foragers is less than F:

$$a + A + (1 - p)GA < F, \tag{4.6}$$

where p is the proportion of producers and G is the size of the group. Reading from the left, the first two terms give the consumption of the producer, who has found the patch, while the third term indicates the amount of food taken by the joining scroungers. If this condition is held, then there is no competition for food within a patch (Barta and Giraldeau 2001). To keep the model simple, we will not take into account the negative effect of overcrowding.

As producers are assumed to find food patches with rate λ, their intake during T time units is:

$$I_p(p) = \lambda T(a + A), \tag{4.7}$$

while the scroungers' intake is:

$$I_s(p) = \lambda TpGA. \tag{4.8}$$

The equilibrium proportion of producers, $\hat{p}$, can be calculated by setting $I_p(p) = I_s(p)$ and solving for p:

$$\hat{p} = \left(1 + \frac{a}{A}\right)\frac{1}{G}. \tag{4.9}$$

A couple of interesting observations can be made on the basis of this simple model. First, the intake of producers is independent of their proportion, and hence the scroungers' proportion. Second, as the intake of producers does not depend on their proportion, the intake of an average individual in an equilibrium group is equal to the intake of a solitary producer. Consequently, the spread of scroungers, if their proportion reaches the equilibrium proportion, does not influence the intensity of resource use. Third, the average per capita intake changes nonlinearly with the proportion of producers. It starts from zero at $p = 0$ and increases through $\lambda T(a + A)$ at $p = \hat{p}$ to have a maximum of $\lambda TGAp_{max}^2$ at $p = p_{max} = (1 + \hat{p})/2$ and finally decreases to $\lambda T(a + A)$ at $p = 1$ (Figure 4.3). Fourth, the equilibrium number of producers, $G\hat{p} = 1 + a/A$, does not depend on group size. Therefore, the same number of producers can support scroungers in groups of widely varying size, up to a limit. This limit, G_{max}, can be estimated as follows. The maximum amount of food taken from a patch is:

$$a + A + \left(1 - p_{max}\right)GA, \tag{4.10}$$

which can be simplified to GAp_{max}. G_{max} is the largest group size for which $G_{max}Ap_{max} < F$.

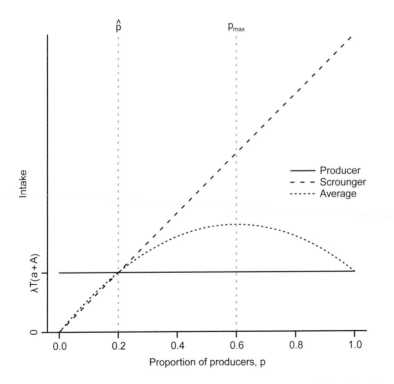

Figure 4.3   The intake of producers and scroungers is shown as the function of proportion of *producers* in a group using ephemeral resource patches (i.e., when patches disappear before the foragers can fully consume them). The intake of an average individual in the group is shown by the dotted line. The vertical dashed gray lines mark the equilibrium proportion of producers ($\hat{p}$) and the proportion of producers when group intake is maximized (p_{max}), respectively.

Discussion

Long-term use of common-pool resources usually requires two types of investment: in the actual withdrawal of resources from the resource system ("appropriation"; Ostrom 1990) as well as in the maintenance of the resource system itself ("provision"; Ostrom 1990). For instance, to obtain wood, an individual must invest in physical labor and tools to harvest timber from a forest as well as plant new seedlings to maintain the forest. Accordingly, exploiters can defect on investors by exploiting both the harvesting and maintenance efforts. Retaining the forest as our analogy, the first way to exploit could mean stealing someone's harvested timber, whereas the second could entail not participating in the planting of new trees. Both ways of exploitation reduce, at the level of individuals, the gain of investors and increase the immediate gain of exploiters (Figure 4.4 and 4.5). As a result, both types of exploitation of others' investment lower the proportion of investors. Nevertheless, at the level of the whole resource system, their effects differ (Figure 4.4 and 4.5).

In this chapter, by reviewing evolutionary game theoretical models of producing and scrounging, I have illustrated how decreasing the proportion of producers (investors) lowers the amount of resources that the group, as a whole, uses (e.g., in the case of exploiting harvesting efforts). If the group overuses its resources, the spread of scroungers (those who exploit the investment of others) can mitigate the environmental problem because fewer investors means that more resources will be left (Figure 4.4). Circumstances like energy reserves or dominance hierarchy facilitate the spread of exploiters whereas aggression and compatibility of producing and scrounging impede it. Kin-related benefits usually increase production, but if producers are able to discriminate between kin and non-kin scroungers, as many bacteria can (Brown, this volume), exploitation increases in kin groups, leading to a lower level of resource use. These results may indicate that there are ways to influence the spread of exploiters, even in humans, and hence alter resource use (e.g., by human groups). This creates, however, a moral dilemma. While moral value is generally not considered when we examine the exploitation of others' efforts in animal groups, it is regarded as bad in human communities. Therefore, to combat resource overuse by humans, it does not seem to be morally acceptable to encourage such exploitation. Nevertheless, these models offer a different interpretation: resource use by the group could be lowered if some of the investors' benefits were redistributed to those who refrain from investing. This could be accomplished, as results of social preference studies in behavioral economics show that humans are willing to sacrifice some of their own benefit if this improves the well-being of others (e.g., Charness and Rabin 2002). Implementation of a suitable (or more equitable) tax system by policy makers might offer such a solution. The effectiveness of this measure, however, needs be very carefully evaluated before any attempt is made to implement it. Finally, I note that these conclusions hold solely for common-pool resource

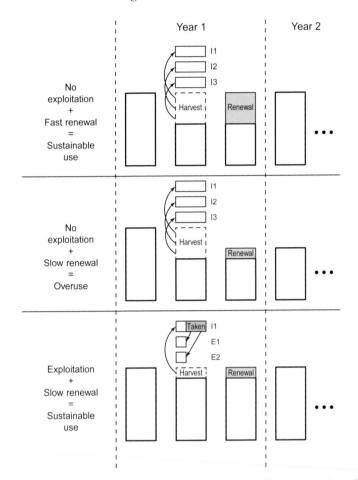

Figure 4.4 Schematic illustration of how exploitation affects harvesting efforts. Available resources are shown by boxes with thick borders. (a) Three individuals (I1, I2, I3) invest in harvesting (symbolized by the curved arrows) part of the resource (box labeled "harvest"). The renewal of the resource (gray box labeled "renewal") is fast enough to replenish the harvested resource (cf. the first box in Year 1 with the box in Year 2). (b) Renewal is slow and cannot cover the resources used by the three investors; the resource is thus overused. (c) Two individuals (E1, E2) exploit the effort of a sole investor (I1) and take (straight arrows) part of the resource harvested by I1 (small gray box labeled "taken"). This results in an equal amount of the resource being consumed and no resource overuse.

systems. In the similar public goods resource system, where ephemerality prevents competition, the spread of exploiters does not effect the resource use by the group, as discussed above.

The exploitation of maintenance efforts differs fundamentally from the exploitation of harvesting efforts (Figure 4.4 and 4.5). Exploitation of maintenance reduces the amount of available resources and can undeniably lead to an

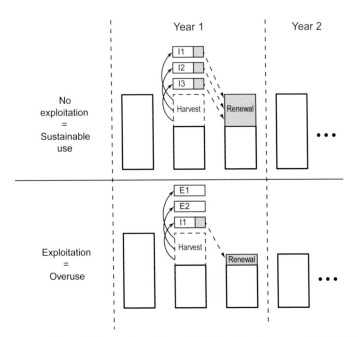

Figure 4.5 Schematic illustration of the exploitation of maintenance efforts. (a) Three investors (I1, I2, I3) invest (straight dashed arrows) part of their harvested resources (small gray boxes) into the renewal of the resource system ("renewal"), which allows the resource system to recover. (b) One investor (I1) invests into the renewal of the resource system while two exploiters do not. This results in the overuse of the resources and a larger gain for the exploiters.

overuse of resources (Figure 4.5). In addition, exploiters always exact a larger gain than investors because they achieve the same from the result of the investors' investment but save the costs of investment (Figure 4.5). Accordingly, there is nothing to prevent the spread of exploiters. As a consequence of these differences, the exploitation of maintenance efforts resembles the prisoner's dilemma game and can easily result in a "tragedy of the commons" outcome (for a more detailed comparison of the two scenarios, see Valone et al., this volume).

An important point neglected by the models reviewed here is density dependence. One could argue that groups which contain many scroungers use resources less efficiently than groups of producers. The remaining surplus, however, would allow more individuals to forage on the area. The increased number of individuals finally results in the same level of resource use as the groups consist of only producers. This seems reasonable, but to clarify how density dependence influences resource use in a social foraging setting we need more detailed PS game models with explicit density dependence included. The situation might be complicated by the fact that the cost and benefit of foraging tactics could change with the age of the individuals, as the state-dependent

PS game suggests. This indicates that we need an age-structured population model to investigate how density dependence affects resource use in social exploitation.

An important result of this review is that exploiters (scroungers) can significantly influence the state of the resource systems. Therefore, their existence needs to be taken into account when managing natural resources. A literature search, however, revealed that the governance of natural resources co-occurs with cheating in only 493 (3.3%) out of over 14,000 articles that presumably address the management of natural resources (Table 4.1). This 3.3%, a negligible interest, indicates that the examination of cheating and defection is largely missing from the studies of governance of natural resources. The models reviewed here as well as the findings of this literature search suggest that management of natural resources could be substantially improved by considering the effects of scroungers (or free riders).

Conclusions

The evolutionary games of producing and scrounging provide an appropriate framework to investigate the effect of exploitation of harvesting effort on resource use by humans. As resource use usually takes place in groups, this kind of exploitation is almost inevitable. A review of current PS models indicates that the spread of scroungers (exploiters) decreases resource use in the population. According to these models, several factors can facilitate as well as impede the spread of scroungers in a group. This indicates that it might be possible to lower resource use in human groups by setting conditions to promote scrounging. The moral dilemma that results from this awaits careful analysis and further investigation.

Table 4.1 Results of an online literature search on the link between governance of natural resources (Search Set #1) and cheating (Search Set #2). The column "Search Term" gives the actual search terms used on Thomson Reuters' Web of Science web site on 02/09/2015. Search Set #3 returned those articles that presumably addressed cheating in resource management.

Search Set	Search Term	Number of Articles
#1	(((ecosystem NEAR/2 service$) OR (natural NEAR/2 capital$) OR (natural NEAR/2 resource$)) AND (governance OR management))	14,721
#2	((free rider$) OR (cheating) OR ((prisoner's OR prisoners) AND dilemma$) OR (social NEAR/2 dilemma$) OR ("tragedy of commons") OR (public good$) OR (scroung*))	74,846
#3	#1 AND #2	493

Acknowledgments

I am grateful to Sam Brown, Juan-Camilo Cardenas, Gareth Dyke, Fred Thomas, Bruce Winterhalder, and two anonymous reviewers for their useful comments on the manuscript. I am especially thankful to my editors Luc-Alain Giraldeau, Philipp Heeb, Michael Kosfeld, and Julia Lupp for their guidance through the writing. The publication was supported by a NKFIH grant (grant no. K112527) and the SROP-4.2.2.B-15/1/KONV-2015-0001 project. The project has been supported by the European Union, cofinanced by the European Social Fund.

5

Common-Pool Resource Management

Insights from Community Forests

Johan A. Oldekop and Reem Hajjar

Abstract

Community-based natural resource management (CBNRM) initiatives aim to link socioeconomic development with sustainable natural resource use and the conservation of biodiversity of natural resources. The principal ethos of CBNRM relies on the concept that rights, responsibilities, and authority for natural resource management decisions should rest, at least in part, with local communities, and there is an increasing recognition among policy makers and practitioners that the decentralization of natural resource management is central to a rights-based sustainable development approach. Although there has been a global push to decentralize natural resource governance over the past two decades, outcomes have been mixed, with many initiatives failing to reach their intended goals of both natural resource conservation and socioeconomic development. Over the past few decades, much research has focused on identifying the kinds of enabling conditions and accompanying institutional arrangements needed to promote collective action (investing) and reduce free riding (exploitation) to bring about more sustainable and equitable management of shared resources. This chapter reviews the theory and conditions thought to aid and allow communities collectively to manage resources more equitably and sustainably. The management of community forests is used to explore current knowledge gaps related to collective resource management and discuss what these gaps represent for sustainable development interventions.

Introduction

Conserving the world's natural resources while ensuring human well-being and socioeconomic development is central for a transition to a more sustainable development path (UN 2015; UNFCCC 2015). Community-based natural resource management (CBNRM), which aims to link socioeconomic development with sustainable natural resource use and conservation, emerged in the 1980s as a way to counter conservation inefficiencies and the negative social outcomes associated with top-down natural resource management initiatives (Batisse 1997). The fundamental principles of CBNRM are centered on the concept that rights, responsibilities, and authority for natural resource management and conservation decisions should rest, at least in part, with local communities. Evidence suggests that local communities are managing an increasing amount of the world's natural resources,[1] and it is now widely accepted that the decentralization of natural resource management is central to a rights-based sustainable development approach (UN 2015).

The theoretical framework underpinning CBNRM, and its justification as a sustainable development and conservation strategy, is largely based on common-pool resource theory. In his seminal article, "The Tragedy of the Commons," Garret Hardin essentially describes an *n*-person cooperative game in which resources "open to all" are inexorably destined to be overexploited: individuals will always "seek to maximize [their] gain" because the consequences of overexploitation "are shared by all" (Hardin 1968:1244). In this context, the tendency would be for all individuals to play the role of "exploiters" until the resource is depleted. This understanding of the consequences of an open-access resource spurred a generation of top-down natural resource conservation policies and interventions (one of his recommendations for overcoming the tragedy), where rules are set from the top-down to limit resource use and punish exploiters (West et al. 2006a).

Hardin's work, however, has been widely criticized for failing to distinguish between "open-access resources," which are devoid of any property rights, and "common-pool resources," where property rights are clearly defined (either formally or informally) and held by a specific set of individuals (e.g., Ostrom 1990). There is substantial evidence that communities with devolved decision-making powers and secure rights to resources will often act collectively to create local institutions (rules, practices, and norms) to manage local resources and avoid tragedy of the commons scenarios. Key to these local institutions is the ability to monitor compliance and administer sanctions against transgressors and free-riding individuals that exploit more than their allocated share of a collective resource. The relationship between cooperating individuals and free riders in common-pool resource systems is, hence, conceptually analogous to other "investor–exploiter" models, such as those found in social foraging scenarios in many animals: "exploiter" individuals or species exploit the investment made by "investor" individuals that either provide or discover new resources. The aim of local institutions in common-pool resource systems is to create social and economic incentives that will shift livelihood strategies within a group of resource users from short-term individual maximization of

---

[1] Tenure Data Tool, Rights and Resources Initiative, http://www.rightsandresources.org/en/resources/tenure-data/tenure-data-tool/ (accessed Aug. 25, 2016).

gains or less collective action (exploiting) to long-term sustainable collective management (investing) (Ostrom 1990).

Beginning in the mid-1980s, a large and rich literature combining elements of political science, behavioral economics, anthropology, and ecology has focused on the kinds of enabling conditions and accompanying institutional arrangements that promote collective action and reduce free riding to bring about more sustainable and equitable management of shared resources (Ostrom 1990; Agrawal 2001; Cox et al. 2010). Understanding the factors and conditions under which communities can encourage shifts from exploitative unsustainable livelihood strategies toward ones in which collective resources are managed more sustainably and equitably is of paramount importance for the design and implementation of better conservation strategies at local and global scales.

In this chapter, we provide a brief overview of the conditions thought to facilitate collective action and effective common-pool resource management. Using the community forest management literature as a case study, we explore current knowledge gaps related to common-pool resource management and discuss what these gaps represent for sustainable development interventions. We focus on community forests for three principal reasons. First, community forests represent one of the most widely researched collectively managed resource systems, with direct measurable links to both local livelihoods and key environmental outcomes (Hajjar et al. 2016). Critically, they share fundamental commonalities with other common-pool resource systems (a clearly defined resource managed by a clearly defined group of users) such as fisheries or irrigation systems, and are influenced by similar internal and external factors (e.g., markets, population dynamics, and local institutions). Thus, key lessons gleaned from community forest management studies will also apply to other common-pool resource systems. Second, community forests are central to national and international sustainable development agendas (UN 2015; UNFCCC 2015), including key global initiatives such as the United Nations Collaborative Programme on Reducing Emissions from Deforestation and Forest Degradation (REDD). Finally, many governments have implemented forest decentralization policies. Communities are thought to manage approximately 15% of forests globally,[2] and thus play a key role in conservation efforts.

Conditions Facilitating Implementation and Longevity of Local Institutions

In her influential book, "Governing the Commons," Elinor Ostrom (1990) exhaustively analyzed different common-pool resource management arrangements

---

[2] Tenure Data Tool, Rights and Resources Initiative, http://www.rightsandresources.org/en/resources/tenure-data/tenure-data-tool/ (accessed Aug. 25, 2016).

around the world and identified a series of principles that facilitate collective action and the implementation and longevity of common-pool resource institutions. Since its publication, this initial list has been revised to include additional factors (e.g., Agrawal 2001; Cox et al. 2010), which collectively focus on:

- the resource system being managed,
- the user group managing the resource system,
- the relationships between the resource system and the managing group,
- the institutional arrangements and their relationship to the resource system, and
- external factors, such as market forces, and higher-level governance arrangements.

Resource System Characteristics

Factors related to the resource system include the size of the resource, with evidence that local institutions are more easily enforced in communities managing smaller-sized resources with well-defined boundaries than larger ones (e.g., Chhatre and Agrawal 2008). Similarly, resources whose availability or quantity is unpredictable in space and time, and difficult to store, or those where the individual resource units are highly mobile are thought to be more difficult to manage and monitor. For example, it is easy to foresee that monitoring the abundance of game species or fish stocks that move in and out of a given collectively managed area might pose significant difficulties (e.g., Jenkins et al. 2011).

User Group Characteristics

There is substantial evidence that leadership and the way in which local leaders dispense sanctions for transgressions and free riding play a vital role in mediating the success of collective action and the outcomes of common-pool resource management (e.g., Kosfeld and Rustagi 2015). In addition, the size of a group is thought to influence a group's ability to forge relationships and trust, and mobilize resources to implement local institutions (e.g., hiring a guard to monitor compliance). Smaller, better-defined groups might be more willing to work together because their shared contributions and collective benefits might be more tangible: if benefits are dispersed over too large a group or if an individual contribution is perceived as inconsequential, the incentive to work collectively diminishes (Poteete and Ostrom 2004). Evidence suggests that the relationship between group size and collective action is, however, complex and nonlinear (e.g., Oldekop et al. 2010). Some studies suggest that larger groups might be able to manage common-pool resources more effectively than smaller ones (Nagendra et al. 2005) but that collective efficiency diminishes as groups get too large; that is, when the costs associated with collective management supersede both individual and collective benefits (Agrawal and Goyal 2001; see also Burton-Chellew et al. and Valone et al., this volume). Similarly, groups which share values and common goals might be more willing to work together (Poteete and Ostrom 2004). However, the relationship between social heterogeneity and collective action is highly complex. Some evidence suggests that greater socioeconomic heterogeneity affects collective decision making negatively (e.g., Balooni et al. 2007), whereas others suggest that the causal relationship between collective action and social heterogeneity flows both ways, and that highly heterogeneous groups might act collectively when faced by a common threat (e.g., Johnson 2001).

Linking Resource System and User Group Characteristics

Several conditions and relationships between common-pool resource users and the resource system are thought to mediate collective action (Agrawal 2001). First, physical proximity between resource users and the resource itself is considered to be crucial because there is an increased cost to monitoring distant resources. Second, levels of resource demand by the user group should be low and changes in demand should be gradual. Communities might be less willing or able to manage vital resources that are in high demand because the immediate individual cost of overexploitation is lower than the perceived benefit gained from collective long-term management. Furthermore, local institutions might not be able to change or adapt fast enough to sudden changes in demand. Third, resources should be perceived as finite, as there is little incentive to manage resources that are not perceived as scarce (Oldekop et al. 2012). Finally, the allocation of benefits and access to resources should be equitable; competition for resources by economically differentiated groups might lead to conflict and hence lower collective action (Poteete and Ostrom 2004).

Institutional Arrangements

The creation and implementation of rules, monitoring protocols, and sanctions in relation to the use of common-pool natural resources is considered key for the promotion of collective action and reduction of free riding necessary for effective common-pool resource management (e.g., Persha et al. 2011). In some instances, these institutions can be based largely on traditional resource use and management systems, as is the case in many indigenous reserves in Latin America (Davis and Wali 1994), whereas in others, institutions are embedded within official management plans that are devised in conjunction with government departments, as is the case in both India and Nepal (Agrawal and Ostrom 2001). Whenever possible, however, resource access and management rules should be locally devised, simple, and easy to understand, enforce, and arbitrate, and they should match natural resource regeneration rates and cycles (Ostrom 1990; Agrawal 2001). For example, the establishment of monitoring protocols to oversee noncompliance and graduated sanctions to punish free riders has been linked to better resource management outcomes (Ghate and Nagendra 2005; Chhatre and Agrawal 2008). Critically, those involved in monitoring should also be locally accountable (Gautam and Shivakoti 2005).

External Factors

Common-pool resources and their users are embedded within highly complex, broad social and ecological systems and governance arrangements (Ostrom 2009). Overall, local common-pool resource management institutions are thought to be more effective when they are supported by broader governance arrangements and not undermined by regional authorities or central governments (Ostrom 1990). Similarly, commodity and labor markets can influence individual and collective livelihood decisions, including the increase of harvesting rates (Oldekop et al. 2013), decisions to emigrate (Uriarte et al. 2012), and changes in the distribution of benefits within local communities. This can reduce collective action, weaken local institutions, and result in negative environmental outcomes.

Over the past decade, the effects of climate change have become of central concern to CBNRM debates, both because the rural poor are likely to experience the greatest disadvantages and because local communities that manage common-pool resources might be able to play critical roles in climate change mitigation efforts, including payment for ecosystem services schemes such as REDD+. However, to date we lack an adequate understanding of the direct effects of climate change (including climate change-related policy changes and interventions) and the impact of private sector investments (e.g., large-scale land transactions) on local communities, collective action, and their abilities to implement local institutions.

Forest Commons as a Way to Understand Common-Pool Resource Management

The literature on community forests provides a useful and policy-relevant case study with which to explore our theoretical and applied understanding of common-pool resource management (e.g., Persha et al. 2011). Many governments have decentralized forest management since the 1980s (Agrawal et al. 2008). In Nepal, for example, current legal rights that permit local communities to take part in forest management are enshrined in the country's Forest Act of 1993 and the Forest Regulations of 1995 (Agrawal and Ostrom 2001), and today there are more than 18,000 community forest user groups managing more than a quarter of Nepal's forests. Similarly, in Mexico, the communal Ejido system—officially introduced as part of agricultural land reforms in the 1930s—has been strengthened through a series of policies since the 1970s, and estimates suggest that local communities manage more than half of the country's forests (Bray et al. 2003). Indeed, conservation and development practitioners have increasingly promoted community forestry initiatives as a way to enhance sustainable forest use, consolidate rights over traditional lands and resources, and reduce rural poverty (Bray et al. 2003; Molnar et al. 2008). Community forest management has become central to global sustainable development initiatives, such as REDD+. The longevity and scale of many community forestry initiatives, and their direct link to environmental and social outcomes and sustainable development policies, have led to a rich body of research focused on trying to understand the factors that lead to successful community forest management outcomes (Hajjar et al. 2016).

Although scores of case studies around the world show that community forestry, collective action, and the implementation of local institutions can potentially improve sustainable forest use and livelihoods (Pagdee et al. 2006; Oldekop et al. 2010; Bowler et al. 2012), outcomes have often been mixed. Many initiatives have failed to reach their intended goals (Edmunds and Wollenberg 2003; Oyono 2005). To date, most studies focusing on social and environmental outcomes of community forest management have concentrated on assessing the effects of institutional arrangements associated with community forests, examining both the effects of tenure, local institutional arrangements, and collective action on livelihoods, forest biodiversity, and carbon storage (e.g., Chhatre and Agrawal 2008; Persha et al. 2011; Newton et al. 2016). Several meta-analyses have aimed to determine factors that lead to community forestry success (Pagdee et al. 2006; Oldekop et al. 2010; Baynes et al. 2015), including a review of the links between land tenure and deforestation (Robinson et al. 2014), and an examination of whether formal community forest management has been more effective than other management arrangements (Bowler et al. 2012).

In comparison to the social conditions and institutional arrangements that can lead to collective action and more efficient local institutional arrangements, our understanding of the role of social, political, economic, and biophysical factors in shaping collective action and community forest outcomes—or indeed, the comparative effect of different kinds of community forest management arrangements—remains very limited (Hajjar et al. 2016). However, elucidating the relationships known to affect livelihood decisions, collective action, and forest dynamics at various scales is key for providing a strong evidence base, which in turn is needed to design and implement better decentralized natural resource management policies.

In their recent systematic review of 735 cases from the peer-reviewed literature on forest commons, Hajjar et al. (2016) evaluated the occurrence with which studies reported information on 53 variables related to user group characteristics and demographic factors, local institutional arrangements and market characteristics, and biophysical characteristics. Their results highlight several important issues and knowledge gaps. First, research on community forest management continues to focus on assessing the role of institutional factors (see Appendix 5.1), despite significant evidence that demographic changes (e.g., migration-led population shifts), market forces, and biophysical factors significantly influence collective action and resource management decisions (Agrawal and Yadama 1997; Agrawal and Chhatre 2006; Uriarte et al. 2012; Oldekop et al. 2013) as well as forest and land cover change dynamics (Geist and Lambin 2002; Rudel et al. 2005; Meyfroidt and Lambin 2011). Second, most studies used qualitative measures to assess the effect that community forest management and collective action have on livelihood outcomes. Furthermore, other critical development outcomes (e.g., food security, which some have suggested could be promoted through community forest management initiatives) have not been given much formal attention. Although these studies have provided valuable insight into the kinds of socioeconomic impacts that community forestry initiatives can have and the types of collective management arrangements and local institutions that drive them, there is an urgent need to complement these studies with quantitative measures using standardized indicators to make comparative assessments of intervention outcomes across sites, and to help establish baselines for longitudinal studies.

Finally, there appears to be a heavy bias in the forest commons literature toward South Asian countries (predominantly India and Nepal). Thus the literature might not be representative of global decentralization and community forestry interventions. The area of forests in Latin America under community control is an order of magnitude larger than in Africa or South Asia (225.75 Mha versus 22.89 Mha and 28.27 Mha, respectively[3]), yet cases from Africa represent a quarter of the reported analyses in the literature, and India and Nepal represent more than half.

Filling in the Knowledge Gaps

Arguably, forest commons represent one of the best-studied common-pool resource systems. The rich literature on community forest management has provided valuable theoretical insight into the social and institutional conditions that promote positive socioeconomic and environmental outcomes of collective action and common-pool resource management arrangements. To date, however, studies on social and environmental outcomes of community forest management have typically only focused on individual case studies and analyzed a limited set of variables at a single point in time. Analyses that focus on general patterns and overall trends at larger geographical scales have either relied on meta-analyses (Pagdee et al. 2006; Oldekop et al. 2010) or relatively small-*n* studies (e.g., Persha et al. 2011). These studies have focused

---

[3] Tenure Data Tool, Rights and Resources Initiative, http://www.rightsandresources.org/en/resources/tenure-data/tenure-data-tool/ (accessed Aug. 25, 2016).

predominantly on evaluating the effect of institutional arrangements on forest outcomes, and relatively little attention has been paid to other potential confounding factors. Currently we know of only two studies that have used a more robust impact evaluation approach to assess deforestation over time (Rasolofoson et al. 2015) and livelihood outcomes (Pailler et al. 2015) of community forest management at the country level, while also controlling for a large set of potentially confounding variables. Thus, understanding of the impacts of decentralization policies remains limited.

National- and regional-level evaluations, such as the ones conducted by Rasolofoson et al. (2015) and Pailler et al. (2015), are critical because they provide general assessments of the effectiveness of policies and interventions, which often operate at large geographical scales. Assessments at these larger scales can highlight significant regional variations and provide critical information about specific enabling conditions and circumstances (social or biophysical) that lead to different outcomes or trade-offs between social and environmental goals of interventions (e.g., Andam et al. 2010). Critically, understanding the overall impacts and enabling conditions along with their nuances is necessary for the design and implementation of more effective policies and interventions.

The increasing ease of using remote sensing tools (e.g., Hansen et al. 2013) has delivered standardized measures with which to assess environmental outcomes in forests over large areas (e.g., Nagendra et al. 2005). However, the collection of socioeconomic data is typically costly and difficult to implement in a sufficiently coordinated fashion to allow the creation of comparative data sets for more robust analyses (Poteete and Ostrom 2008). For example, the International Forestry Resources and Institutions research program has been collecting data on community forests, using a standardized methodology, in 11 countries for over twenty years but has only been able to start revisiting some of these sites in the past few years, which will no doubt provide valuable insight into the long-term outcomes of community forest management.

No central databases on community forests currently exist, and there is a clear need to devise better data collection programs and assessment protocols to evaluate outcomes of CBNRM initiatives (Baylis et al. 2016; Oldekop et al. 2016). Global, publically available data sets at high spatial resolutions are becoming more widely available, and can provide standardized, consistent, and longitudinal information on a host of social and environmental variables, including agricultural suitability[4] and climate data (e.g., Hijmans et al. 2005), socioeconomic data and measures of poverty,[5] travel times to population centers

---

[4] METI-NASA, http://asterweb.jpl.nasa.gov/gdem.asp (accessed Aug. 26, 2016).

[5] World Bank. Living Standards Measurement Survey. http://microdata.worldbank.org/index.php/catalog/lsms (accessed Aug. 18, 2016).

and distance to markets,[6] subnational-level administrative areas,[7] forest cover (e.g., Hansen et al. 2013), and biomass (e.g., Avitabile 2016). While these data sets can provide valuable information on variables that may likely influence the outcome of community forest management initiatives, efforts to combine them with fine-scale data on local institutional arrangements and collective action are still lacking. Better coordination among academics is clearly needed; however, governments, donor agencies, and implementation agencies also need to invest and implement better monitoring and evaluation protocols and ensure that the resultant data sets are publically available. Collectively, these efforts could provide both better understanding of successes, failures, and trade-offs between social and environmental outcomes. This is clearly needed for both better theory and the design and implementation of more effective policies and interventions. Critically, developing better analytical protocols for the analysis of community forests can also provide valuable frameworks for the assessment of other common-pool resource systems, including community-based fisheries management and collective pasture management systems.

Conclusion

Communities that have decision-making powers and secure rights to the natural resources upon which they depend will often act collectively to design and implement local institutions to manage them. These local institutions, in turn, operate as social and economic incentives that aim to shift livelihood strategies within a group of users managing a common-pool resource from short-term overexploitation (exploiters) to long-term sustainable collective management (investors). Over the past 25 years, a rich and diverse scientific literature has focused on trying to understand the social, economic, and environmental enabling conditions that promote collective action and the implementation of long-lasting institutions, and whether these lead to better local socioeconomic and environmental outcomes. A review of the community forest management literature, which provides a large and useful case study example of common-pool resource management, shows that studies to date have predominantly focused on understanding the local institutional arrangements and how these are linked to sustainable outcomes for both people and forests. Despite being acknowledged as important, much less attention has been given to demographic factors (such as population shifts, e.g., outmigration), market forces, or biophysical factors that impact forest dynamics. Understanding how these factors influence the ability of communities collectively to manage forests and

---

[6] Global Accessibility, https://people.hofstra.edu/geotrans/eng/ch1en/conc1en/global_accessibility.html (accessed Aug. 18, 2016).

[7] GADM database of global administrative areas, version 2.8, http://gadm.org (accessed Aug. 18, 2016).

implement local institutions is a crucial first step in figuring out how to impact positive change in terms of socioeconomic and environmental outcomes.

From both a theoretical and applied empirical perspective, solely focusing on institutional arrangements to explain "investor–exploiter" relationships in CBNRM and the effect of these relationships on social and environmental outcomes is insufficient, because such efforts fail to account for important contextual factors. We argue that the increased availability of publically available environmental and socioeconomic data, such as remote sensing and national census data sets, can provide novel theoretical and empirical insights on the effectiveness of CBNRM initiatives. Although the integration of such data sets is not straightforward, the opportunities which they provide for broader regional- and national-level studies could yield key insights on the factors that drive variation in social and environmental outcomes. This, in turn, is key for understanding the socioeconomic and biophysical factors that drive sustainable livelihood shifts in common-pool resource management systems and is of paramount importance if we are to design better targeted policies and interventions that support them.

Acknowledgments

We thank Luc-Alain Giraldeau and Philipp Heeb for constructive comments on a previous version of this chapter. JAO was supported by a European Union FP7 Marie Curie Research fellowship.

Appendix 5.1 (next page) Data map indicating the incidence of reported variables in 735 case studies of the community forest management literature (dark gray = recorded data, pale gray = missing data). Variables are thematically grouped and data rows are grouped by country highlighting those countries with ten cases or more. The community forest literature has predominantly focused on institutional factors and environmental outcomes. Variables associated with population dynamics (density, change, and migration), market forces, and biophysical factors feature less prominently. Reproduced from Hajjar et al. (2016).

94

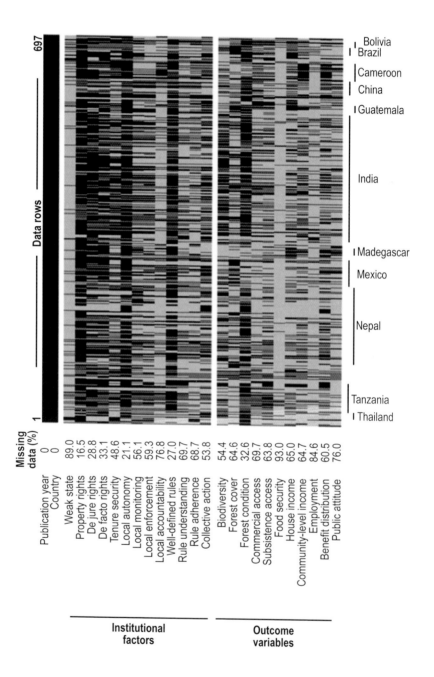

95

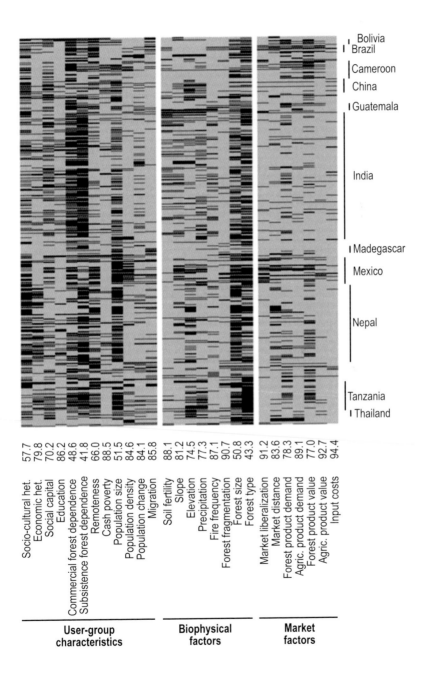

6

Governance of Renewable Resources

Insights from Game Theory

Thomas J. Valone, Zoltán Barta, Jan Börner,
Juan-Camilo Cardenas, Luc-Alain Giraldeau,
Hanna Kokko, Johan A. Oldekop, Daniel Pauly,
Devesh Rustagi, and William J. Sutherland

Abstract

Renewable resources have the potential to be used in a sustainable manner but typically are not, often due to the existence of exploiters or free riders. This chapter analyzes free-riding behavior using the prisoner's dilemma-based public goods model and the producer–scrounger model. Overuse of renewable resources is examined under four investor–exploiter scenarios that are derived from modifications of the classic producer–scrounger model, and which vary in the degree of excludability of a discovered resource and in the cost of adopting each strategy. Two important factors are found to reduce overuse: when a finder's advantage can be created for investors, and when the costs of playing exploiter are increased relative to the costs of playing investor. Applying the investor–exploiter model to a fisheries scenario, discussion follows on how interventions designed to reduce overuse may be consistent with the existence of a finder's advantage. A variety of existing interventions can be seen as increasing the costs of adopting the exploiter strategy.

Introduction

Biological resources are able to renew themselves via reproduction and thus can be potentially harvested in a sustainable fashion. As is widely accepted,

Group photos (top left to bottom right) Thomas Valone, Zoltán Barta, Luc-Alain Giraldeau, Jan Börner, Daniel Pauly, Devesh Rustagi, Johan Oldekop, Hanna Kokko, Thomas Valone, Daniel Pauly, William Sutherland and Luc-Alain Giraldeau, Jan Börner, Luc-Alain Giraldeau, Juan-Camilo Cardenas, Johan Oldekop, William Sutherland, Devesh Rustagi, Zoltán Barta, Hanna Kokko, Juan-Camilo Cardenas, Devesh Rustagi

biological resources are currently overused, and this situation constitutes a major global conservation problem (Diamond 1989; Pauly et al. 2002). The use of biological resources by humans greatly exceeds that of animal herbivores, piscivores, or carnivores (Darimont 2015). Well-documented declines in tropical forests, bushmeat, and fish stocks are linked to human exploitation and have resulted in a reduction of harvest rates and benefits (Pauly et al. 2002). Halting the degradation and overuse of biological resources, while maintaining and enhancing human well-being, are critical steps that must be achieved if humankind is to transition toward a more sustainable society. In the classical "tragedy of the commons" scenario popularized by Hardin (1968), individuals who manage open-access common resources (i.e., a resource without a defined set of users or property rights) behave according to their own self-interest, thereby depleting a common resource used by all. Such overuse (or "inefficiency" in economic terms) can be reduced through various governance rules designed to curtail resource use by groups and individuals. However, to implement effective resource management interventions, the decision-making processes reached at the individual and group levels (and the factors which influence these decisions) must be clearly understood.

Problems of resource use are not unique to humans. A fascinating example of unfortunate resource management is provided by the Amazon molly (*Poecilia formosa*), a fish species that does *not* get its name by living in the Amazon—its distribution actually spans areas in Texas and Mexico—but rather because of a similarity between aspects of its reproductive system and the Amazon women of Greek mythology. According to legend, these women killed all of their male offspring and thus needed to travel to other villages to secure fertilization. Amazon mollies do not kill their male offspring; they simply do not produce any, because all their eggs develop into daughters, which are clones of their mother. Their form of asexual reproduction is rare and is termed gynogenesis (or sperm-dependent parthenogenesis): eggs still need to come into contact with sperm before they begin developing. This is a vestigial trait of their past history as a sexual species. Molecular evidence shows that the Amazon molly is the result of two sexual molly species hybridizing. All of the genes that the sperm contains are actively rejected by the egg.

This situation of mothers needing sperm while producing only daughters creates a problem of sperm supply. The species can only exist in the presence of at least one "sperm donor" species: another species of mollies that have retained males. Amazon molly females look very similar to the females of the sexual species, so males may have a hard time discriminating. As a result, the system can work, but only for a while. Amazons, by avoiding the need to produce males, are twice as fecund as their sexual sperm provider, because only females directly produce offspring. Therefore, Amazons avoid the so-called twofold cost of sex and, over time, ecologically outcompete their sperm-providing sexual species causing their extinction, which in turn, can lead to the Amazons' own extinction. To our knowledge, this is the only fish species for which there is a published mathematical proof that they should not exist (Kiester et al. 1981)! To account for their existence, Kiester and colleagues highlighted that other factors (e.g., spatial structure that allows the sexual species to persist) must be added to the basic population dynamic model to explain species coexistence (Kokko et al. 2008). Humans, of course, would rather not live in a spatial mosaic of local extinctions and subsequent recolonizations of their resources. The development of governance structures or rules (e.g., quotas on harvests) would thus be desirable to reduce the possibility of the tragedy of the commons.

One factor commonly associated with overuse of renewable resources is the existence of free riders (i.e., exploiters that take advantage of the investment of others). One example of free riding can be found in the creation of quotas to reduce overuse. Here, a governance rule imposes restraint on how much can be harvested. Individuals who engage in this restraint make an investment: their actions generate higher resource densities, which makes harvesting more profitable for all. Their investment, however, becomes vulnerable to exploitation by individuals who disregard the quotas, as these free riders enrich themselves by harvesting beyond set quotas (also known as "quota busting"). Such free-riding behavior is common yet detrimental to all, since it reduces the density of the resources generated by the investors' restraint (Munro 1979).

To explore further such exploitation strategies as free riding, and their impact on resource use, let us look at two different conceptual approaches: the public goods model based on the prisoner's dilemma (Axelrod and Hamilton 1981) and a modified version of the producer–scrounger model (Barnard and Sibly 1981). Our aim is to examine how these different approaches are related and can be linked to models of resource use, thus providing insight into more effective governance rules.

Framing of the Problem in a General Framework

Empirical studies of resource use, both in animal and human societies, often reveal the coexistence of at least two strategies in a population engaged in using resources. One strategy can be generally described as *investor* because it consists of investing in making a resource available. This strategy has received various appellations, such as *producing* or *cooperating*. The term "investor" is meant to apply whenever actions, as a net effect, lead to maintenance or increase in the resource of interest (possibly over time) in a dynamic setting. Investors increase the availability of a resource for a group (a resource has been newly discovered and made available for exploitation) but there is a cost associated with this behavior. An investment is not necessarily equal to "acting." Restraint from a behavior can still make an individual an investor in our sense, both when it requires installing new technology (e.g., using better equipment to harvest) and when it does not (e.g., by reducing harvesting effort); the latter captures the idea of choosing to invest in our children's future at a direct cost to ourselves (Sumaila and Walters 2005).

The second strategy, which coexists with the investor, is that of the *exploiter* and has been variously labeled as *scrounging, defecting, free riding, kleptoparasitizing, piracy*, or *stealing*. The exploiter does not pay the costs of generating new resources or maintaining them, but instead attempts to usurp some of the resources produced or maintained by others. It is worth noting that categorizing individual behavior along the investor–exploiter axis does not mean that this is the only trait axis along which individuals can vary. There could be individuals who are neutral along the investor–exploiter axis yet differ in other details. Variations outside this axis can then place an individual, as a net effect, as an investor or an exploiter.

Public goods models have frequently been used to understand the existence of investor- and exploiter-like strategies within populations of animals and human societies. These models can vary in the number of players, strategies, and the parameters and properties of the payoff functions. The simplest of games is the canonical version of the prisoner's dilemma. Here, investing yields a benefit, b, at a cost, c. Players face the payoffs detailed in Figure 6.1. Such configuration will comply with the key properties of prisoner's dilemma games if $b > c > 0$. A more general version of the prisoner's dilemma is presented in Figure 6.2, which shows the four outcomes.

When more than two players are involved, several models extend the possibilities of other similar collective action problems. Public goods games, for instance, involve a number of players that must decide on how much to invest from a private asset into a public fund that produces benefits to all players involved (Archetti and Scheuring 2012). The private cost of investing, however, is greater than the benefit the investor receives from her own investment in the public good if others also do not invest. On the other hand, if all players invest in the public fund, the sum of all payoffs to all players increases. This n-person prisoner's dilemma game creates a situation where the dominant strategy (or Nash equilibrium) is not to invest (cooperate) but rather to exploit (defect), creating a situation in which universal defection yields the worst possible outcome for the group. The socially optimum solution is for all players to cooperate (invest) (Archetti and Scheuring 2012).

	Cooperate	Defect
Cooperate	$b-c$, $b-c$	$-c$, b
Defect	b , $-c$	$0, 0$

Figure 6.1   Canonical version of the prisoner's dilemma, showing the payoffs of two opponents, who can either cooperate (invest) or defect (exploit). The first entry in each cell gives the payoff of the row individual; the second entry shows the payoff of the column individual: b, benefit; c, cost; $b > c > 0$.

	Cooperate	Defect
Cooperate	R, R	S, T
Defect	T, S	P, P

Figure 6.2 General version of the prisoner's dilemma game. The first entry in each cell gives the payoff of the row individual; the second entry shows the payoff of the column individual: T, temptation from defecting; R, reward from cooperating; S, sucker's payoff; P, punishment payoff; $T > R > P > S$, $2R > S + T$.

Producer–scrounger models, by contrast, envision producers (investors) that discover a food resource and scroungers (exploiters) that exploit some fraction of the discovered food. These strategies are mutually exclusive in the sense that an individual can only adopt one or the other at any given moment. Players, however, can switch between strategies sequentially, so it is important to bear in mind that "producer" and "scrounger" do not refer to individuals but rather to strategies that individuals may adopt at a specific point in time. An individual that adopts the producer strategy searches or otherwise invests in making a good available at a personal cost of c. The good, once encountered, has some value b ($b > c$). In certain situations, as we see below, the model predicts some stable equilibrium frequency of producer and scrounger strategies within the population.

Both the prisoner's dilemma and producer–scrounger models envision a form of "social parasitism" in which some individuals can benefit from the costly behavior of others. Here, we seek to frame the problem in the context of resource use, identify how these models are related, and outline future work. We suggest that, to a large extent, the choice of conceptual approach to be adopted in analyzing renewable resource governance depends on the nature of the resources that need to be governed. These resource characteristics will determine when some individuals will behave in ways that benefit others, while others do not. Economists refer to these resource characteristics as the *rivalry* and *exclusivity* dimensions. Rivalry means that use of a resource by one makes it unavailable to others; this is what ecologists call depletion. The exclusivity dimension gives the extent to which others can be completely excluded from its use, something behavioral ecologists would refer to as defendability or despotism. Resources that are both nonexclusive and without rivalry are termed pure public goods. This means that once generated, everyone can benefit from the resource, irrespective of the strategy being used. At the other extreme, a resource with maximum rivalry and high exclusivity profits only the individual that made it available and/or its usurper.

A Simple Social Resource Use Game

Below, we examine four scenarios in a simple social resource use game that involves two strategies: investor and exploiter. These strategies are mutually exclusive in the sense that an individual can only adopt one or the other at any given moment. However, players can sequentially (and rapidly) switch between strategies. An individual that adopts the investor strategy searches or otherwise invests in making a good available at a personal cost of c ($c > 0$). The good once encountered has some value b ($b > c$), and we consider a group, G, of individuals. The proportion of exploiters in the group is p. We assume that resources are rivalrous (depletable) and consider four scenarios where exclusivity of resource use differs.

Scenario 1: Entire Group Exploits All Resources Made Available by an Investor

In this case, all individuals can use the resource made available by an investor. The investor's gain is:

$$W_I(p) = \frac{(1-p)Gb}{G} - c = (1-p)b - c. \tag{6.1}$$

The factor $(1-p)$ is important because it means that gains can only arise when individuals adopt the investor ("cooperative") strategy; the greater the number of individuals who do so in a population, the greater the common good produced for all.

An exploiter's gain is:

$$W_E(p) = \frac{[(1-p)Gb]}{G} = (1-p)b. \tag{6.2}$$

It follows that $W_E(p) > W_I(p)$ for all values of p. As a result, the expected Nash equilibrium solution, or the evolutionarily stable solution in evolutionary biology, is noncooperative: the exploiter strategy spreads over the entire population even if this results in zero gain for all individuals in the population (Figure 6.3). This resembles the solution of the n-person prisoner dilemma game and

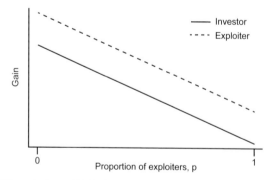

Figure 6.3   The gain (payoffs) for investors (producers) and exploiters (scroungers) as a function of the frequency of exploiters in a population. In this case, the payoffs to the exploiter are always higher than to the investor. The model thus predicts a population of all exploiters, which corresponds to the n-person prisoner's dilemma solution.

is also the case of the pure public goods game. Empirical work with humans in the laboratory has shown that under the above conditions, individuals do converge toward a universal exploiter (free-riding) strategy when no communication is allowed among group members (Ledyard 1995).

Scenario 2: All Individuals Exploit Part of the Resources Made Available by an Investor

Now let us allow the investor to secure part ε of the produced resource for itself and refer to this as the *finder's advantage*. This situation may arise if agents are, for instance, involved in finding or making a good available that is not immediately available to everybody in the group. This delay makes it possible for the investor to secure part of the good before the arrival of others. Imagine, for instance, the discovery of a gold mine on public land; this would have a special effect on the individual that discovers it, as a result of its own producer behavior, and a secondary effect on all individuals. Now only the surplus, $(1-\varepsilon)b$, is available to everyone else, including the finding investor and all other individuals playing investor. We assume limited excludability, $\varepsilon < 1$, otherwise the resource would be completely exploited by its investor.

When $\varepsilon < 1$, the investor's gain, which comes from its advantage as well as from joining the discoveries of other investors, is:

$$W_I(p) = \varepsilon b - c + \frac{(1-p)G(1-\varepsilon)b}{G} = \varepsilon b - c + (1-p)(1-\varepsilon)b. \quad (6.3)$$

The exploiter's gain is given by:

$$W_E(p) = \frac{\left[(1-p)G(1-\varepsilon)b\right]}{G} = (1-p)(1-\varepsilon)b. \quad (6.4)$$

When the level of excludability is low (i.e., the size of the finder's advantage is $\varepsilon < c/b$), then the equilibrium solution is that all should be exploiters (Figure 6.4). Note: Scenario 1 is a special case of Scenario 2, with $\varepsilon = 0$. On the other hand, if the level of excludability is high enough, $\varepsilon > c/b$, then investors do better than exploiters independently of their proportion; hence the stable solution will be for the investor strategy to spread until all play investor.

In a social foraging context, this scenario corresponds to the information-sharing model of Clark and Mangel (1986). Next we investigate cases when the surplus is available to only parts of the group, depending on the individual's strategy.

Scenario 3: The Surplus Is Shared by the Exploiters Alone

In this case, an investor only gains from resources produced by itself and exploiters can only access the surplus. This could happen if exploiters posed a

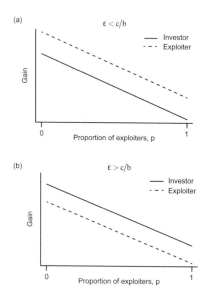

Figure 6.4   The gain (payoffs) for investors (producers) and exploiters (scroungers) as a function of the frequency of exploiters (scroungers) in a population. (a) When ε < c/b, exploiter always has higher gain than investor that corresponds to the n-person prisoners dilemma solution. (b) In contrast, when ε > c/b, investor always obtains a higher gain than exploiter, and the model predicts a population of all investor. Finder's advantage is ε, the cost of investing is c, and b is the value of the resource made available by the investor.

serious threat to investors such that upon the arrival of exploiters, investors always left the resource they created or discovered; for example, a scenario in which farmers produce food that is then stolen by bandits. In this case, the investor's gain is:

$$W_I(p) = \varepsilon b - c. \tag{6.5}$$

The exploiter's gain is:

$$W_E(p) = \frac{\left[(1-p)G(1-\varepsilon)b\right]}{(pG)} = \frac{(1-p)(1-\varepsilon)b}{p}. \tag{6.6}$$

The equilibrium proportion of exploiter, p_e, is:

$$p_e = 1 - \frac{\varepsilon b - c}{b - c}. \tag{6.7}$$

If ε < c/b then p_e = 1 such that no investors remain in the population and hence we have the n-person prisoner dilemma solution; that is, the resource is not provided (Figure 6.5). However, for the situation where the finder's advantage

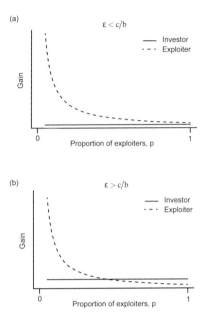

Figure 6.5   The gain (payoffs) for investor (producer) and exploiter (scrounger) as a function of the frequency of exploiter (scrounger) in a population. (a) When $\varepsilon < c/b$, exploiter always has higher gain than investor and the model predicts the n-person prisoner's dilemma-like solution. (b) In contrast, when $c/b < \varepsilon < 1$, investor obtains higher gain than exploiter if there are few individuals playing investor in the group and, vice versa, exploiter gains more than investor if exploiter is rare. Consequently, the model predicts a stable mixture of investor and exploiter. Finder's advantage is ε, the cost of investing is c, and b is the value of the resource made available by the investor.

is of intermediate value (i.e., when $c/b < \varepsilon < 1$, then $0 < p_e < 1$), we expect an equilibrium containing a mixture of investor and exploiter strategies; a mixed evolutionarily stable solution.

Scenario 4: An Investor Gains Its Advantage from a Discovered Patch and Shares the Rest with All Exploiters

The last scenario corresponds to the classic producer–scrounger game, where investors are the producers and exploiters are the scroungers. The producer's gain is:

$$W_P(p) = \varepsilon b - c + \frac{(1-\varepsilon)b}{1+pG}.$$ (6.8)

The scrounger's gain is:

$$W_s(p) = (1-p)G\frac{(1-\varepsilon)b}{1+pG}. \tag{6.9}$$

Again we have an equilibrium mixture of investor and exploiter strategies (Figure 6.6) and the equilibrium proportion of producer is:

$$p_e = 1 - \frac{\varepsilon b - c}{b - c} - \frac{1}{G}. \tag{6.10}$$

Summary

The four scenarios illustrate that this investor–exploiter model predicts different outcomes depending on how resources are shared among the group members (degree of excludability). In turn, this is influenced by particular conditions of the environment (e.g., size of the finder's advantage, threats posed

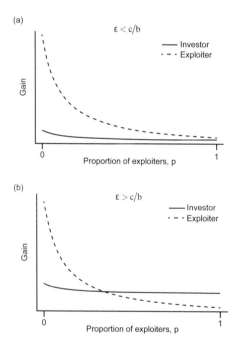

Figure 6.6   The gain (payoffs) for investor (producer) and exploiter (scrounger) as a function of the frequency of exploiter in a population. (a) When $\varepsilon < c/b$, exploiter always has higher rewards than investor, and the model predicts the *n*-person prisoner's dilemma solution. (b) In contrast, when $\varepsilon > c/b$, investor and exploiter have equal gain at an intermediate frequency of exploiter, a mixed evolutionarily stable solution. Finder's advantage is ε, the cost of investing is c, and b is the value of the resource made available by the investor.

by exploiters). If none of the group members can be excluded from using the produced resources (scenario 1) or an investor cannot secure enough of the produced resources to cover its cost of investing ($\varepsilon < c/b$), then the model predicts the same noncooperative, all exploiter evolutionarily stable solution or Nash equilibrium as an n-person prisoner's dilemma game would do. If all individuals have access to the produced resource but the investor retains a large enough part of the resource ($\varepsilon > c/b$), then the evolutionarily stable solution or Nash equilibrium is the cooperative solution: everybody plays investor (scenario 2). In the remaining cases, a mixture of investor and exploiter is predicted. A crucial aspect of scenarios 3 and 4 is that investors have no access to resources produced by other investors. These cases correspond most closely to the classic producer–scrounger game model.

Next we investigate how these different scenarios affect resource use and management by applying this general model to a specific example: resource management of a renewable biological resource such as a fishery.

Basic Conceptual Fisheries Model

A great deal of modeling has been directed at understanding how harvesting efforts affect total fish stock biomass (Clark 1990; Hilborn and Walters 1992). This can be summarized in the basic conceptual model of fisheries as viewed from the perspective of both the entire fishery and an individual fisher or company (Figure 6.7). From the perspective of the entire fishery, the value of the resource extracted (fish) increases first with great extraction effort (e.g., number of fishing days or number of vessels), peaks at a maximum (the maximum sustainable yield, MSY), and then declines as more extractive effort is applied, because the stock has declined. Since the extraction process (e.g., fishing) is associated with a cost, maximum profits or "rent" (i.e., the maximum difference between the revenue and the cost curve: maximum economic yield, or MEY) are achieved at a level of effort lower than required to generate maximum sustainable yield (MSY).

While MEY or MSY may be viewed as reasonable targets or limits for society as a whole, they require some degree of regulation (e.g., quotas on total harvest). This is true because individuals using the resource maximize their net benefit by harvesting until their benefit equals their costs at the equilibrium point (EQ) (Figure 6.8). In practice, and especially in fisheries, this equilibrium point is far to the right of MSY and thus stocks are overharvested to a low level.

In well-regulated fisheries, a quota at or below MEY or MSY is allocated among agents: fishing at this level of effort will maintain the biological and/or economic productivity of the fishery. The key problem illustrated in Figures 6.7 and 6.8 is that society benefits most when the fishery is extracting MSY or MEY, but individuals benefit more, in the short term, by exploiting at a higher rate.

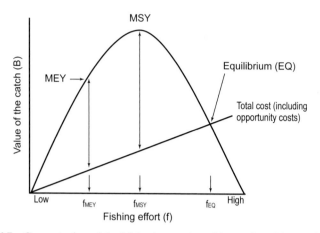

Figure 6.7 Conceptual model of fisheries as viewed by society. The total catch increases when fishing efforts increase until the maximum economic yield (MEY) or maximum sustainable yield (MSY) is reached. Afterward, total catch declines because of continued use, toward B = 0, when the stock is exhausted. However, equilibrium occurs when individuals maximize their net benefit (their total costs = B): here, effort is relatively high and stock biomass is low.

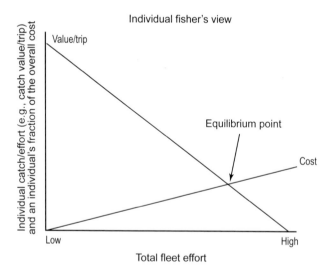

Figure 6.8 Conceptual model of fisheries as viewed by an individual agent. An individual fisher, fleet, or firm's benefits (value of the catch) and individual's fraction of the overall costs are shown on the y-axis as a function of the total aggregate catch effort. The individual fishers' catch per unit of effort, and hence their gross return, declines with aggregate effort using the stock biomass, and trends toward zero, as total fleet effort increases because the stock biomass becomes zero. Equilibrium occurs when individual value/trip equals the cost at relatively high total fleet effort.

Application of Investor–Exploiter Models to Fisheries

Can the investor–exploiter modeling approach offer insight into fisheries management? To approach this question we must establish correspondences between the investor–exploiter game and fisheries. One tool commonly used in fisheries is the establishment of quotas that limit total catch effort. Quotas are divided between agents (individuals and firms) and are designed to maintain stock biomass near MSY. To bring this into the investor–exploiter framework, we assume that individuals who agree to abide by such a system are investors; that is, they invest in restraint to maintain the stock levels. Exploiters, on the other hand, are agents who do not invest in restraint while exploiting the stock. They can be framed as engaging in "quota busting" by harvesting beyond a specified quota.

Therefore, the investor–exploiter model is applicable at the level of an individual agent (whether fisher or firm). First consider a typical public goods fishery (Figure 6.9). The x-axis represents the proportion or frequency of individuals that engage in the exploiter strategy on that play of the game. The x-axis also represents stock biomass (with higher biomass values to the left). Any increase in the frequency of exploiters will reduce total stock biomass because they exploit the stock without investing in restraint; they do not abide by the quota agreement that attempts to maintain stock biomass at MSY. Investors exhibit restraint and thus abide by the quota system. Their payoff is highest when all invest (Point A: no exploiters), and investor payoff declines as the frequency of the exploiter strategy increases because stock levels are reduced. Exploiter always obtains higher payoffs than investor because it harvests more fish, but its payoffs also decline with increasing frequency of the exploiters strategy (again because total stock biomass is reduced). The equilibrium is a population where all individuals engage in the exploiter strategy (Point B) leading to the *n*-person prisoner's dilemma solution and overexploitation (as seen in scenario 1).

How can we avoid the *n*-person prisoner's dilemma solution that results in overexploitation of resources? Examination of scenarios 2–4 reveals that one important factor involves the existence of a finder's advantage which produces an extra benefit to investors and is unavailable to the exploiter strategy. Are there conditions in fisheries where the resource is rivalrous and excludable?

One possibility lays in the creation of a benefit that would be exclusive to the investor strategy. Such a mechanism would lead to a decline in the exploiter payoff curve (Figure 6.9) because exploiters would have access to less resources. A finder's advantage could be accomplished through various certification programs (e.g., the Marine Stewardship Council), where agents engaged in the investor strategy invest in abiding by various regulations that limit their catch. All other fishers whose investment in restraint is unknown or undocumented would be playing the exploiter strategy, because they are likely not abiding by quota agreements or are otherwise fishing unsustainably. One could argue that

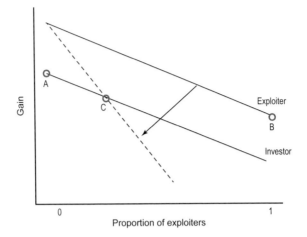

Figure 6.9 The conventional fisheries model of Figure 6.8 presented as a classic form of the investor–exploiter game (solid lines). Investors exhibit restraint (e.g., use large mesh nets, limit days fishing, or avoid Marine Protected Areas) and this results in maximum economic yield; exploiters do not invest in such measures and, as a result, always achieve a higher catch (higher gain). The game theory solution is for all to play exploiter (Point B). Note, however, that the highest overall catch is when all play investor (Point A), which exceeds the catch at the equilibrium of all playing exploiter. One way to avoid a group of all exploiters is when the negative frequency dependence of payoffs to the exploiter strategy is stronger (has a more negative slope) than for the investor strategy. This can occur if investor obtains a special protected share of the resource that exploiters can never access (i.e., a finder's advantage). Another way is for exploiter to pay a strategy-specific cost. For instance, if there is punishment for widespread failure to invest in restraint, then actions such as fines or boycotts force the gains obtained by exploiter to decline (dashed line) faster than the gain to exploiter, such that there is now an equilibrium balance between investor and exploiter where the lines cross (Point C).

the certification provides a finder's advantage to the investor and thus could make the payoff lines cross at some intermediate frequency of investor and exploiter. In theory, certification programs allow consumers to select between cheaper products from fisheries with less oversight and more costly products caught according to sustainability criteria. However, in practice, due to the incentive structure, it can be true that certified fisheries are unsustainable and/ or that sustainable fisheries cannot afford certification (e.g., Jacquet and Pauly 2008; Christian et al. 2013).

Another possibility lays in increasing the costs of adopting the exploiter strategy, which again should lead to a reduction in the exploiters' payoff curve (Figure 6.9). A significant body of work has focused on elucidating the conditions and collective arrangements (institutions) under which groups of individuals can manage renewable resources more sustainably (i.e., use the investor strategy rather than exploiter).

Policies (interventions) are the mechanisms by which individuals can be encouraged to play investor rather than exploiter (free rider). These often involve increasing the costs of adopting the exploiter strategy and include, for example, command-and-control, conditional payments, access rights, and punishment (see Figure 6.1 and Appendix 6.1). Interventions alter behavior and lead to the reallocation of resources among actors (Ostrom 1990). Thus it makes sense to categorize interventions in terms of how they intend to affect behavior (Börner and Vosti 2013). For example, positive incentives (e.g., subsidies or payments for environmental services) can reduce biological resource overuse by transferring financial resources in a society from beneficiaries of external biological resource services to owners of these resources. Negative incentives or disincentives (e.g., taxes) can have the same effect on the biological resource, by transferring financial resources from biological resource users to the beneficiaries of external service beneficiaries.

While the standard environmental policy model would predict the same biological resource outcome independent of the policy instrument, we know from behavioral economics that the direction of the financial resource transfer can have different effects on the response of resource users and external beneficiaries, such as through motivational crowding (Bowles and Polania-Reyes 2012). Another mode of intervention that can be labeled "enablement" addresses the conditions affecting collecting behavior. This could include redistribution of property rights (e.g., land reform), education, technology development, or the management of beliefs and norms, such that cooperative (investor) behavior prevails among a defined group of resource users. Examples of interventions in each category are provided in Table 6.1.

Recent work has focused on social norms and the role of leadership in affecting behavior. Social norms are rules of behavior that we expect others in our group to follow; when they do not, we expect such deviant behavior to be punished or shamed. Young (2008) defines social norms as *"customary rules of behavior that coordinate our interactions with others."* The expectation of

Table 6.1 Examples of different interventions intended to affect behavior.

	Intervention/ Type	Intended impact channel	Example
Incentives	Conditional transfers	Compensation for opportunity costs of resource maintenance	Payments for environmental services
Disincentives	Command and control	Increase the costs of resource degradation	Resource use limits subject to fines or punishment
Enablement	Establishment of property rights	Encourage long-term investments in resource maintenance	Decentralization
	Management of norms and beliefs	Behavioral change	Nudging, awareness-raising

shunning by others in the group would sustain the compliance of the group, and therefore it could provide an endogenous institution to solve social dilemmas like the tragedy of the commons. Norms, however, do not always align social and individual interests.

In the case of resource use, norms could regulate (without the need for an external authority) actions related to technology (e.g., fishing gear), efforts used to extract resources, and the sharing of costs or benefits. Baland et al. (2006) argue that social norms can shape behavior by either limiting the action set of the players or by changing the preferences of the players. Maintaining a social norm (making sure that everyone is doing his or her fair share of making the norm to be preserved) is individually costly but benefits all. Norms are sustained through shame, guilt, and embarrassment (Elster 1989) and can emerge and be evolutionarily stable for solving common-pool resource dilemmas (Crawford and Ostrom 1995; Sethi and Somanathan 1996). These norms could be used, then, to encourage resource users to switch from scrounging–exploiting to producing–investing by transforming the relative payoffs to the player from the two strategies.

Finally, the role of leadership in galvanizing group cooperation figures prominently in social sciences. One way that leaders can resolve cooperation dilemmas is to lead by example, whereby leaders contribute first and encourage group members to follow (Gaechter and Renner 2014). Another way is to act as a punishment authority. However, evidence from development economics suggests that leadership can also have a negative effect (Bardhan and Mookherjee 2002). In fact, Kosfeld and Rustagi (2015) show leaders of groups engaged in forest commons management vary in their motivation to punish, and this has implications for the performance of groups in managing their forest commons. Leaders who emphasize equality and efficiency see positive forest outcomes. Antisocial leaders, who punish indiscriminately, see relatively negative forest outcomes. In addition, experiments in the field conducted in Mali show that leaders are more effective in inviting community members to contribute in public goods games (Alzua et al. 2014). These results highlight the importance of leaders in collective action and, more generally, the idiosyncratic but powerful roles that leaders may play, leading to substantial variation in group cooperation outcomes, by encouraging members to shift toward investing strategies.

Future Directions

The investor–exploiter model provides an alternative view of renewable resource management. Its main application here has been to fisheries, but we assume that other renewable resource management problems—those that involve excludability of the resource, and affect both investors and exploiters in resource maintenance efforts—could potentially exhibit similar properties.

Further work should extend the model to actual systems. This will involve the difficult task of identifying individuals engaged in playing investor from those engaged in exploitation. One possibility might occur in tuna fisheries. Tuna fishing is frequently conducted with the aid of fish aggregating devices (FAD), which in the past consisted of palm fronds and currently are made of metal and/or concrete. Fads take advantage of the propensity of tuna to swim under floating objects (Floyd and Pauly 1984). Investing in FADs can be very costly, but the agent that does so can expect to be able to harvest the fish that they aggregate. Other agents who do not invest in FADs can, if they are able to locate them, raid the FADs as thieves or scroungers. In this case, producer involves investing effort in the construction of FADs. By doing so, producers have a higher likelihood of harvest from them (because they know their exact locations), and thus they stand to gain from a finder's advantage.

A second possible example involves the Philippines, where legislation authorizing the creation of marine reserves by coastal municipalities has led to hundreds of such reserves. Local fishers play investor when they invest in restraint by not fishing within the reserve. However, they can also obtain enhanced fishing in areas adjacent to the reserve due to their enhanced knowledge of local conditions and the likely movement of fish out of the reserve to adjacent waters, due to their high population sizes inside reserves. This enhanced fishing can perhaps be considered as an investor's (finder's) advantage. Exploiter then consists of fishing in the reserve, a behavior that is sometimes adopted by the fishers of neighboring municipalities, or by industrial vessels. Again, it is uncertain whether this fits the investor–exploiter game. If it did, it would result in the stable equilibrium of both investor and exploiter strategists, and stock biomass levels would remain higher than if the n-person prisoner's dilemma solution existed (Pollnac et al. 2001).

In principle, the investor–exploiter approach might also help us address a long-standing problem in resource extraction by quantifying illegal harvest. If a resource extraction system has an internal equilibrium with investor and exploiter strategists, the model predicts that the payoffs to those playing investor are equal to those playing exploiter. Consider a fishery system with certification in place. The catch of investor strategists that are certified can be readily quantified because they gain from transparency (e.g., amount of fish biomass harvested). From this quantity, given the assumed equilibrium frequency of investor and exploiter, one could potentially generate an estimate of the (illegal) harvest by the exploiter strategists if the total number of fishers is known.

Resource management is complex, in part because both the resource and the exploiting agents are heterogeneous and operate on different scales. The majority of successful examples of collective resource management come from individual case studies of local resource use settings (Pretty 2003). While these studies highlight factors that have led to better outcomes, far too little is known about how sustainable biological resource governance can emerge at national, regional, and global scales.

Conclusions

Economists view renewable resource management as a problem of efficiency, because of the existence of free riding, and biologists use the producer–scrounger game to study how individuals divide up into investors and exploiters, depending on environmental conditions or species attributes. These two views have necessarily evolved for specific purposes, but they do have commonalities. In the underlying theoretical models, for example, changes in resource attributes affect behavioral responses in similar ways, and different interventions (policies) can affect the level of human cooperation much as changes in costs and benefits affect the frequency of producers and scroungers in a group.

To achieve interdisciplinary synergy, the alignment of terminology and theoretical concepts poses a major challenge. Our sustained interest in understanding the curious dynamics of animal species, such as the Amazon mollies, which mismanage their resources, may offer sufficient proof of the potential benefits to be gained by pushing toward more integration among disciplinary theoretical frameworks in biological resource management and beyond. We hope that this contribution serves as an initial step in this direction.

Acknowledgments

We thank Julia Lupp and Marina Turner for their skills and unyielding encouragement. We also thank the staff of the Ernst Strüngmann Forum for the outstanding level of support provided during our meeting. This contribution would not have been possible without the generous support of the Ernst Strüngmann Foundation: thank you. Finally, we would like to thank Jennifer Jacquet and Rashid Sumaila for reading and commenting on a draft of this manuscript. WJS is funded by Arcadia. JAO was supported by a European Union FP7 Marie Curie international outgoing fellowship (FORCONEPAL). ZB was supported by a NKFIH grant (grant no. K112527).

Appendix 6.1: Punishment

In human societies, evidence suggests that cooperation can be enhanced by disciplining free riders, either through punishment or by forming rules and enforcing them (Fehr and Schmidt 1999). A study by Ostrom and Nagendra (2006) of multiple forests owned by states, communities, or private firms revealed that regardless of the property rights regime, forests were in better shape when the users were involved in the design, monitoring, and sanctioning of the rules that governed them. A growing number of studies have now identified coercive behaviors in biological systems as an important aspect of cooperation. In biology, punishment refers to cases where the act of punishment reduces the punisher's fitness at least initially, i.e., without taking into account any changes in its partner's future behavior (Raihani et al. 2012). If this was the end of the story, it would be hard to understand how punishing behavior could evolve. One solution appears to be that punishment can become *self-serving* if the changes of partner behavior or identity make it beneficial for the punisher. It may also be that the initial reduction in fitness—the cost of punishing—does not really arise in the first place.

A series of experiments in coral reef fish help to illustrate these points. A fish, the scalefin anthias (*Pseudanthias squamipinnis*), has a problem in the presence of other fishes, such as sabertooth blennies (*Petroscirtes* spp.) which attack anthias and other fish victims from behind. The bitten fish then chases the blenny, but is this "retaliation" costly? An initial energetic cost is presumably present, but this behavior has been shown to decrease the probability of future attacks by the same fish. Here, a public goods situation appears to be created for the entire shoal, because chasing also increases the probability that the attacker next time chooses to target another fish species. Finally, an experiment showed that blennies appear to be able to discriminate between fish that do and do not chase them, even if they look alike (to us) (Bshary and Bshary 2010), making "self-serving punishment" the best explanation overall. This, however, should not be taken to suggest that cost-free or even low-cost punishment is necessarily the norm in nonhuman societies (Pollock et al. 2004).

Exploitation in Public Health

7

Does Social Exploitation within Pathogen Populations Pose an Opportunity for Novel Therapeutic Approaches?

Sam P. Brown

Abstract

Bacterial virulence (damage to host) is often cooperative, with individual cells paying costs to promote collective exploitation. This chapter reviews how cooperative virulence traits offer novel therapeutic avenues involving either the genetic introduction or chemical induction of "cheats" that can socially exploit cooperative wild-type infection. Issues of efficacy and evolutionary robustness are discussed, and evidence of an evolutionarily robust therapeutic that targets bacterial social behaviors is highlighted.

Social Evolution in Microbes and Pathogens

Cooperative Microbes

Many species of microbes are now recognized to be highly social, with individual microbes working within large collectives to engineer their environments to allow further growth and dispersal. Collective behaviors include investments in extracellular foraging, collective shelter, dispersal, signal-mediated coordination, and the production of chemical and biological weapons (Crespi 2001; West et al. 2007a; Brown et al. 2009a; McNally and Brown 2015).

Investment in these collectively beneficial behaviors often bears an individual cost (typically the cost of producing and secreting extracellular molecules) and therefore poses a basic social dilemma: Why should a cost be paid to benefit others? More directly, how can cooperative producer lineages survive in competition with nonproducer "cheat" lineages, which take the benefits but do not pay the costs? In the terminology that has been adapted in this book, cooperative producer cells can also be viewed as *investors* in microbial common goods (producing and consuming common goods), while nonproducer cheats can be viewed as *exploiters* of these goods (less production and/or more consumption) (see, e.g., Brown and Taylor 2010).

Solutions to the classic social dilemma posed by cooperative behaviors are various in detail but can all be placed within two classes: nepotism and self-interest (West et al. 2007b). Self-interest may favor cooperative investments if the returns on investment to the actor outweigh the costs of investment. Nepotism (or kin selection) may favor cooperative investments if the return on investment falls preferentially to other individuals carrying copies of the allele(s) that code for the cooperative trait. Both classes of solution can be captured by Hamilton's rule (Hamilton 1964), reviewed in detail by Gardner et al. (2011).

Pathogenic (and opportunistically pathogenic) microbes can also be highly social. For instance, pathogenic and nonpathogenic strains of *Escherichia coli* show broadly equivalent levels of investment in secreted molecules (Nogueira et al. 2009). However, these collectively beneficial secreted factors are now likely to come at the expense of the host and are typically labeled "virulence factors" in the biomedical and microbiology literatures (Allen et al. 2014). For a recent review of microbial sociality in infections, see Leggett et al. (2014).

Pseudomonas aeruginosa as a Model System

The opportunistic pathogen *P. aeruginosa* is a leading experimental model system for bacterial sociality. *P. aeruginosa* is an impressive environmental generalist, able to grow in diverse soil, aquatic, and host environments. As an opportunistic pathogen, it displays an extraordinary host range, from protists to plants to animals. In humans, *P. aeruginosa* infects burns, cuts, catheters, implants and, most notoriously, the lungs of cystic fibrosis patients. This incredible environmental range is also associated with an extensive battery of secreted factors (McNally et al. 2014).

Iron Scavenging by Secreted Siderophore Molecules

One of the first bacterial social traits (and noted virulence factor) to receive extensive experimental attention was the collective capture of limiting iron via secretion of the *P. aeruginosa* siderophore, pyoverdine. Secreted pyoverdine molecules bind to insoluble ferric iron(III), with the resulting pyoverdine-iron complex now accessible to uptake by any cell that expresses an appropriate receptor. Exploiting the ability to construct "cheat" strains by knocking out the ability to produce siderophores (but leaving uptake intact), Griffin et al. (2004) were able to grow cooperative (producer) strains with nonproducer cheats under different metapopulation structures, to test basic social evolution theory on the conditions maintaining costly cooperative traits. They demonstrated that the cooperative lineage was able to survive and approach fixation across a metapopulation only when relatedness was high (each subpopulation founded by a single clone) and the scale of competition was global (more productive patches were able to export their productivity via greater propagule release into a general migrant pool) (Figure 7.1). Since this work, many studies have extended our understanding of siderophore-mediated social interactions (discussed further below, under the section on Chemical Cheat Therapy and Antivirulence Drugs). For a critical exchange of views on the merits of social interpretations of siderophore production, see Zhang and Rainey (2013), Kummerli and Ross-Gillespie (2014), and Rainey et al. (2014).

Extracellular Proteins and Quorum-Sensing Control

In addition to siderophores and other small molecules, bacteria invest up to 3% of their genome coding for proteins that are directly secreted from the cell, with secretome-rich genera including *Bacillus* and *Staphylococcus* (McNally et al. 2014). While intracellular proteins can be efficiently recycled, secreted proteins are likely to be lost to the environment and therefore have imposed selection for the use of cheap amino acids in their construction (Nogueira et al. 2009). Cost management can also be seen in many bacterial species through the use of complex regulatory circuits to control investment in secreted proteins,

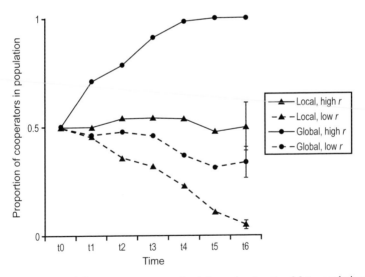

Figure 7.1 Bacterial cooperator versus cheat dynamics *in vitro*. Metapopulation dynamics of cooperators (siderophore producers) versus cheats (nonproducers) as a function of relatedness (high r = single clone founder per subpopulation; low r = two clone founders), and scale of competition (local = all subpopulations contribute equally to migrant pool; global = subpopulations contribute in proportion to their productivity). From Griffin et al. (2004), reprinted with permission.

in particular via the widespread use of quorum sensing to control secretome expression (Popat et al. 2015).

Quorum sensing is a form of cell-to-cell communication in bacteria, mediated by diffusible signal molecules (Rutherford and Bassler 2012; Schuster et al. 2013). The accumulation of high signal levels (due to high densities and/ or low environmental removal rates) then triggers the expression of secreted proteins in *P. aeruginosa* and other quorum-sensing species (Popat et al. 2015). Diggle et al. (2007) presented the first experimental analysis of the social dilemmas posed by signal-mediated control of a costly cooperative trait, again using *P. aeruginosa* as a model system. Using isogenic wild-type (cooperator) and signal-deficient (cheat) strains, Diggle et al. demonstrated that while signal cheats showed attenuated growth in monoculture (as their signaling defects resulted in a loss of cooperative digestive enzyme production), they increase in frequency in mixed culture due to their ability to benefit from extracellular protein degradation by the wild type.

Rumbaugh et al. (2009) subsequently demonstrated similar results *in vivo*, in a mouse model. The signal cheat strains (both "deaf" and "mute" mutants) grew poorly in monoculture and showed attenuated virulence to their host, compared to wild-type infections. However, wild-type coinfections and cheat mutants showed a rapid enrichment of the "cheat" lineages from their intially rare state , and a concurrent attenuation of host mortality (Figure 7.2).

Figure 7.2 provides a clear illustration of the cooperative nature of bacterial virulence, as cooperative investments among pathogen individuals increases exploitation of the host. In the following discussion, I review the potential to control cooperative bacterial infections, either through the introduction of genetic "cheats" or through the chemical induction of a phenotypic cheat state.

Controlling Infections with Genetically Engineered Cheats

The results from Rumbaugh et al. (2009) immediately point toward a novel therapeutic intervention: the presence of initially rare "cheats" (*lasI* or *lasR* mutants) attenuated the infection (Figure 7.2), suggesting that we can reduce virulence simply by adding "cheat" genotypes to a pathogen population to undermine the collective production of host-damaging virulence factors (Brown et al. 2009b).

The notion of "cheat therapy" (Brown et al. 2009b) depends on several complementary processes:

1. Cheats can be genetically engineered to display minimal virulence, in particular through the deletion of secreted virulence factors.
2. The loss of individually costly yet collectively beneficial virulence factors (e.g., siderophores, secreted enzymes, toxins) implies that a relatively small inoculum of cheats will increase in frequency within the

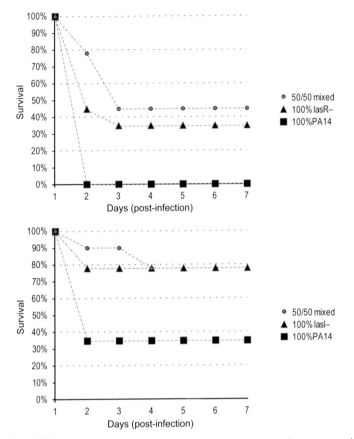

Figure 7.2   Signaling cheats attenuate virulence in mice. PA14 is a cooperative *P. aeruginosa* wild-type strain, with intact quorum-sensing control of secreted virulence factors. *lasR–* is a "deaf" mutant, unable to respond to the primary signal molecule in *P. aeruginosa*. *lasI–* is a "mute" mutant that is unable to produce the primary signal molecule in *P. aeruginosa*. Data from Rumbaugh et al. (2009).

site of infection and therefore attenuate the infection (Rumbaugh et al. 2009; Figure 7.2). For similar amplification kinetics, see phage therapy (Levin and Bull 2004).

3. As the introduced cheat lineage enriches, the total population density is predicted to decline due to a "tragedy of the commons," because fewer individuals contribute to the collective good.

4. The infection can then be fully controlled so long as the introduced cheat is sensitive to an appropriate antibiotic (the "Trojan cheat" strategy, adding a "recall" function to the introduced strain).

While attractive in theory, the potential efficacy of cheat therapy is likely to be limited by a number of processes that can block the ability of cheats to exploit cooperators. Below I briefly review four key obstacles: within-host spatial structure, regulatory control of secreted factors, infectious cooperation (horizontal gene transfer), and concurrent strong selective forces.

Within-Host Spatial Structure

In the idealized *in vitro* setting of cooperator–cheat competition (Griffin et al 2004), cheats and cooperators encounter each other in a well-mixed, shaken flask. This well-mixed regime maximizes the ability of cheats to exploit the cooperative activities of the wild type, favoring the local (within flask) enrichment of cheats. As hosts are clearly not well-mixed flasks, it is no longer evident that cheats will be enriched within a host, due to spatial segregation of the two lineages (increasing the average distance between cooperator and cheat cells). The results of Rumbaugh et al. (2009) illustrate *in vivo* that cheater lineages can increase in frequency. However, this occurred under the specific condition of joint inoculation: the cheat and cooperative lineages were mixed and injected together, and thus likely ended up in a shared within-host environment permitting social exploitation by the cheats. Subsequent sequential inoculation experiments did not show the same enrichment effect, most likely due to sequestration of the initial cooperator lineage away from the subsequent cheater inoculation (Griffin, pers. comm.).

In terms of the potential limitation on cheater invasion posed by within-host spatial structure, it has been suggested that in infection contexts with high spatial structuring, invasion of a therapeutic lineage could be enhanced by the addition of "spiteful" anticompetitor traits, such as the production of bacteriocins and antibiotics by the therapeutic strain (Brown et al. 2009b). The use of anticompetitor adaptations thins the distinction between cheat therapy and the broader concept of competition therapies, such as fecal microbiota transplants to treat *Clostridium difficile* infections (Gough et al. 2011).

Regulatory Control of Cooperative Virulence Traits

The motor to cheater invasion in the models of Brown et al. (2009b) and related theory are the costs of cooperation paid by the wild type: the higher the cost, the faster the invasion by cheats. However, as mentioned briefly above, microbes display multiple, complex adaptations to limit the effective costs of cooperation, for instance, using the cheapest amino acid building blocks when constructing secreted proteins (Nogueira et al. 2009) and complex regulatory rules to limit the potential for exploitation (Kummerli and Brown 2010; Xavier et al. 2011; Allen et al. 2016). The use of regulatory control can ensure that cooperative investments are limited to initial and punctual "start-up costs," which once paid can remove any phenotypic difference and therefore selective differential between cooperator and cheat genotypes (Kummerli and Brown 2010). Regulatory control can also limit investments to environments where costs are negligible (Xavier et al. 2011) or where cooperators are enriched; that is, bacteria can implement simple reciprocity rules of "cooperate only with cooperators"(Allen et al. 2016).

Mobile Genetic Elements and Infectious Cooperation

Bacteria, like any other cellular life form, are vulnerable to molecular parasites or "mobile genetic elements" (MGEs), such as plasmids and phages. Strikingly, as well as damaging their hosts, bacterial MGEs can also confer novel and cooperative phenotypes (Smith 2001), with secretome genes enriched on MGEs versus the chromosome (Nogueria et al. 2009). The presence of cooperative genes on MGEs poses a problem for cheat therapy, and for cheat invasion more generally, as cheat lineages can now become infected with cooperative behavior via horizontal gene transfer, protecting the cooperative phenotype from local extinction (Smith 2001; Nogueira et al. 2009; Dimitriu et al. 2014).

Strong Nonsocial Selection and Local Adaption

The ability of cheats to invade a resident population of cooperators hinges on "all else being equal." In theoretical and experimental treatments, care is taken to ensure that the two strains are isogenic, and thus equally adapted to the experimental conditions or host environment. In practice, a resident wild-type population is likely to be better adapted to the local host environment than any introduced strain. If not, then both strains are likely to be undergoing strong nonsocial selection that is likely to overwhelm any social selection mediated by differences in secreted factor production.

Morgan et al. (2012) demonstrated the potential for strong nonsocial selection to swamp social selection, in an experiment which pitted cooperators (siderphore producers) against cheats in an environment to which neither strain was well adapted. They found that the invasion of rare cheats could be halted and even reversed due to the greater evolvability of the larger cooperative lineage. For example, under strong phage selection, the more numerically dominant cooperator lineage is more likely to experience a beneficial phage resistance mutation. As this rare resistance mutation sweeps, it will then carry the (hitch-hiking) cooperative allele to fixation and exclude the rare cheat lineage.

Chemical Cheat Therapy and Antivirulence Drugs

In addition to introducing genetically engineered nonproducer cheats, the phenotypic state of nonproducers can also be induced chemically via an array of drugs that turn off secreted factors or limit their extracellular function.

The quest for drugs to turn off secreted virulence factors has become a major theme in medical microbiology due to the growing urgency of the antibiotic resistance crisis. "Antisecretion" drugs form a part of a broader theme of "antivirulence" drugs, which aim to disarm rather than kill or cripple our bacterial pathogens. The goal is that by blocking the production of toxins, exoenzymes, and other virulence factors, the pathogen will either return to a commensal state or be more readily cleared by host immune responses (Allen et al. 2014).

Because antivirulence drugs do not directly kill or cripple their bacterial targets, it has been argued that antivirulence drugs confer little or no selection for resistance (Clatworthy et al. 2007; Rasko and Sperandio 2010). The claim of no resistance was first dealt an apparent blow by Maeda et al. (2012), who demonstrated, using a transposon mutant library screen, that several mechanisms of resistance could be found in a novel quorum-sensing interference drug, and that these mutants could be enriched in a specific selective environment. In a recent review, I argue that in light of our experience with antibiotic resistance, mechanisms of resistance will inevitably exist and that the critical question for antivirulence drugs is whether they will increase in frequency under the action of antivirulence drug selection (Allen et al. 2014). Here I will briefly outline the major predictions on the direction of selection as a function of class of virulence factor targeted.

Redundant Virulence Factors and Evolutionarily Robust Drugs

In one scenario, we predict that antivirulence drugs will directly select *against* any resistant mutants, simply because the virulence factors being turned off by the drug are of no benefit to the pathogen's growth or survival within the host (Allen et al. 2014; Figure 7.3a). Under this scenario, any mutant that can restore expression of the virulence factor in the presence of the drug will simply pay the costs of expression without any benefits and be outcompeted. In other words, there is a coincidence of interests between the patient and the pathogen. Both gain by chemically turning off an inappropriate virulence factor.

The broader question becomes: Why would any organism carry a trait that is purely redundant, providing no benefit (direct or indirect) to the individual expressing the trait? Here the answer lies in an understanding of the ecology of bacterial pathogens. The great majority of bacterial pathogens are opportunistically pathogenic in humans, in the sense that their major mode of replication is not in the sites of human disease but instead in some commensal compartment of humans (e.g., *Streptococcus pneumonia*), or some environmental reservoir (e.g., *P. aeruginosa*). What biomedical researchers refer to as "virulence factors" are potentially shaped by distinct functions in these diverse environments (Brown et al. 2012). Due to a lack of broader ecological study of bacterial opportunistic pathogens, we currently lack clear examples of redundant virulence factors, but Allen et al. (2014) propose some candidates.

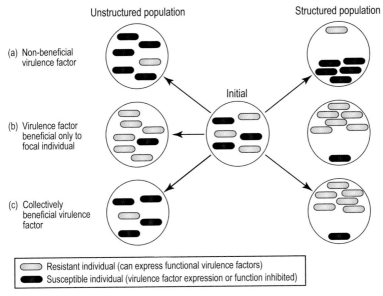

Unstructured population   Structured population

(a) Non-beneficial virulence factor

Initial

(b) Virulence factor beneficial only to focal individual

(c) Collectively beneficial virulence factor

Resistant individual (can express functional virulence factors)
Susceptible individual (virulence factor expression or function inhibited)

Figure 7.3   Predicted direction of selection on resistance to antivirulence therapeutics. Central circle: an initial population of resistant (gray) and susceptible (black) cells. Left circles: the impact of selection in a well-mixed environment. Right circles: the impact of selection in a structured environment. (a) Virulence factors are redundant at the site of infection; (b) individually beneficial virulence factors; (c) collectively beneficial (cooperative) virulence factors. From Allen et al. (2014), reprinted with permission.

Directly Beneficial Virulence Factors and Evolution-Prone Therapeutics

In the second and opposing scenario, the expression of virulence factors is directly coupled to the growth and survival of a focal cell in the site of infection. In this scenario, much like for antibiotics, any cell that has innovated a mechanism of resistance will experience a direct reproductive benefit (Figure 7.3b). Allen et al. (2014) outline several example traits that fall into this class.

Collectively Beneficial Virulence Factors and Mixing-Contingent Therapeutics

The final scenario addresses the set of virulence factors that generate collective benefits. What is the fate of a rare resistant mutant that is able to restore expression of an individually costly yet collectively beneficial trait? As sketched in Figure 7.3c, the answer depends on the degree of spatial mixing of the wild-type (now a phenotypic cheat) and resistant (cooperator) cells. When the drug-targeted population is well mixed, then any drug-resistant cells will produce, at a cost, the cooperative phenotype, yet the benefits will be shared with the drug-sensitive wild-type population, thus selecting against the resistant mutants (Mellbye and Schuster 2011; Allen et al. 2014). If, however, the resistant and sensitive lineages are growing in a sufficiently spatially structured manner, the direction of selection can reverse if mutant cooperator cells are able to benefit preferentially neighboring mutant cooperators (Figure 7.3c; Allen et al. 2014).

A Candidate Evolutionarily Robust Drug with a "Regulatory Trap"

Here I will present evidence for evolutionarily robust control of *P. aeruginosa*, by targeting the collective virulence trait of siderophore production. As discussed above, siderophores are secreted iron-scavenging molecules that bind to and recover insoluble and sequestered iron. Ross-Gillespie et al. (2014) demonstrated that this iron-scavenging ability can be blocked via the addition of low doses of gallium(III) salts. Gallium, like iron, is a transition metal and binds with even greater affinity to extracellular siderophores. We were thus able to use low (sub-cytotoxic) doses of gallium to titrate out the functionality of secreted siderophores and attenuate *P. aeruginosa* infections *in vivo* (Ross-Gillespie et al. 2014). Using a wax moth larvae infection system, we compared survival curves following infection with wild-type *P. aeruginosa,* a siderophore knockout mutant, and wild-type plus chemical (gallium) suppression of siderophore functioning. Strikingly we found that the antivirulence drug is even more effective than the genetic manipulation in reducing mortality rate. In other experiments, we have outlined a likely mechanism for this effect, based on the regulatory response to iron starvation (Ross-Gillespie et al. 2014). The genetic mutant is unable to use siderophores to scavenge for iron and is also relieved of the costs of siderophore production. The gallium-treated bacteria are, however, limited in their ability to scavenge iron. In addition, they pay the costs of continued siderophore production. What is more, under intermediate gallium dosing, bacteria respond to growing iron limitation by increasing their production of siderophores, and thus increasing their costs to a debilitating extent.

The *in vivo* treatment results illustrate that the efficacy of gallium treatment can exceed that of the genetic knockout due to a regulatory trap imposing additional expression costs on the wild type (Ross-Gillespie et al. 2014). How robust is this treatment to evolutionary responses to the drug? The models underlying Figure 7.3 predict that mutants which can restore collective iron scavenging will be selected against if the environment is sufficiently well mixed. We tested this prediction under iron-limited *in vitro* conditions, using a simple serial transfer experimental evolution design, and found support for gallium being evolutionarily robust. Specifically, we found that the degree of growth control imposed by gallium treatment did not significantly decline across the course of 12 days of evolution. In contrast, a range of antibiotic treatments exerting similar degrees of control at day one all displayed significant failure within 12 days, including multidrug treatment (Ross-Gillespie et al. 2014).

Perspectives

The infection-attenuating impacts of genetic and chemically induced cheats suggest that there is some hope of a positive answer to the question posed in this chapter's title: Does social exploitation within pathogen populations pose an opportunity for novel therapeutic approaches? The relatively brief and *in vitro* experimental evolution conducted by Ross-Gillespie et al. (2014) offers further hope that targeting cooperative bacterial traits can lessen or even reverse selection for resistance to these novel therapeutics (Figure 7.3). While these results are encouraging and definitely merit further investigation, a number of concerns remain to be explored.

In terms of efficacy and combination therapy, the first requirement for any new anti-infective agent is that it works at least as well as current treatments. In the case of improvements on antibiotics, most antivirulence drugs are currently stuck in something of a gray area: they are considerably better than antibiotics at treating antibiotic-resistant infections but fall short of the levels of efficacy of antibiotics which treat sensitive bacteria. In the gallium treatment example discussed above, this lack of impressive treatment is apparent (i.e., no lives of wax-moth larvae were saved), although it should be noted that this was a model of an acute infection following injection of *P. aeruginosa* into the hemolymph, and clearance is challenging in this context.

One potential route to improve the efficacy of antivirulence drugs is to couple them with antibiotics in multidrug treatments, because in some scenarios, synergies have been observed (Hentzer et al. 2003). However, by improving clearance, the use of combination therapies will also potentially increase selection for resistance to one or both of the constituent ingredients (Allen et al. 2014).

In closing, in this chapter I have focused on the potential risk of the evolution of resistance and made the claim that in some scenarios this risk will be small. This does not imply, however, that treatment is therefore safe and risk free. Two other classes of ecological and evolutionary risk must also be considered: the epidemiological risk of increasing prevalence following treatment and the evolutionary risk of increased intrinsic virulence. For further discussion of these additional risks, see Vale et al. (2014) and Allen et al. (2014).

8

The Economics of Public Health in the Presence of Exploitation Strategies

Markus Herrmann

Abstract

Public health faces novel challenges because of rising bacterial resistance to antibiotics and the possible fatal spread of highly communicable viral infections. Multiple economic agents interact in the use and provision of anti-infective drugs, without accounting for their impact on others, and this gives rise to positive and negative externalities. Furthermore, anti-infective drugs may be linked on the supply side depending on the particular epidemiological context. A common example is that of antibiotic treatment effectiveness, which can be lost over time, affecting various antibiotics belonging to the same antibiotic family. This chapter describes how conflicting private objectives among economic agents may lead to exploitation strategies that lower the overall social welfare. Important open research questions are highlighted and various possible public policies addressed that can help address the problem of antimicrobial resistance.

Introduction

In developed countries, noncommunicable diseases associated with lifestyle change constitute a primary focus in public health. With the notable exception of HIV and avian influenza, communicable diseases receive less attention. This focus reflects the demographical and epidemiological change that has occurred in developed economies over the last decades, and has increased the demand for treatment of chronic diseases and conditions (e.g., high cholesterol, blood pressure, erectile dysfunction) as well as surgery interventions among the elderly (e.g., prostheses, heart stents). On the supply side, the pharmaceutical industry has kept pace with this evolution (for an empirical analysis, see Pammolli et al. 2011). The development of new drugs and treatments has ensured high, stable revenues while products are patented.

The emphasis in research and development on chronic diseases, however, is about to change drastically. More importance will have to be allocated to the analysis of communicable diseases, even in developed economies, because of the rise of antimicrobial resistance coupled with the potential for new outbreaks of highly lethal viral infections—infections previously thought to have been eliminated by former vaccination programs, but which could reemerge to challenge public health in hospital and outpatient settings worldwide (WHO 2014).

Antibiotics are used intensively today to cure possibly fatal communicable bacterial infections among humans and to prevent infections during surgical interventions. Indeed, they have become a major component of modern human medicine and have contributed to the continuous increase in the welfare of humankind. Furthermore, antibiotics are used intensively to cure and prevent infections in animals (see, e.g., Singer et al. 2003). However, from an epidemiological point of view, it is well understood that intensive antibiotic *use* increases antibiotic *resistance*. Much effort is needed to draw a complete picture of the incentives that are driving economic agents (patients, physicians, pharmacists, hospitals, pharmaceutical firms, and health agencies) in their use of antibiotics.

Economic agents that use or supply drugs, such as antibiotics or vaccines, do not generally account for the potential benefits or costs to third parties. In economic jargon, these benefits and costs are termed "externalities," as the economic agent that causes them does not internalize their effect on others. On the demand side, the use of an antibiotic or vaccine lowers the prevalence of infection in the short and long term; this is an example of a positive externality that is beneficial to society. By contrast, the implied increase in bacterial resistance in relation with antibiotic use in the future implies a negative externality for the individual and society. On the supply side, efforts to preserve antibiotic effectiveness made by economic agents (e.g., a hospital, pharmaceutical producer, or country) may be obliterated by another supplier who does not engage in any preservation efforts, and hence causes important negative externalities.

The potential for free riding on another economic agent's investment—to preserve antibiotic effectiveness or to lower the prevalence of a viral infection via vaccination—is akin to the free riding and the tragedy of the commons (Hardin 1968) that can be observed in exploitation processes of more "classical" natural resources (e.g., open-access halieutic resources or oil fields). Laxminarayan and Brown (2001) were among the first to note the parallel between classical natural resources and those that can be defined in a public health context, such as antibiotic treatment effectiveness. Indeed, using an antibiotic may decrease the "level" of antibiotic effectiveness, such as a classical halieutic resource. When there is no possibility to prevent others from accessing such a resource pool, we can use the economic terminology of a common-pool resource. The particularity of resources related to public health, as described above, is that the free riding may affect both the demand and supply sides, while it only affects the supply side in relation to the more classical resources. Indeed, demand for oil (e.g., for transportation or heating), for instance, remains unaffected when oil resource pools are connected on the supply side, whereas the demand for an antibiotic should reflect the quality of the drug, which is intrinsically related to the level of antibiotic effectiveness available in the pool.

A further difference between the resources analyzed here and classical examples (e.g., halieutic resources and oil) resides in the subsequent development of backstop resources or techniques. As the stock of wild halieutic resources and oil fields diminish, the price of the natural resource should rise over time, allowing aquaculture and solar energy, for instance, to become eventually a competitive backstop technique. In the case of antibiotics, however, no new antibiotic class or technique seems to be able to substitute easily for the decreasing effectiveness of current antibiotic treatment, as was done in the past. New, alternative drugs may, in the future, be able to address genetically engineered or chemically induced bacterial strains that exploit the "social" interaction between bacteria to decrease bacterial fitness (see Brown, this volume). However, since these drugs are only at the beginning of development, their effectiveness remains highly uncertain.

As a consequence, preserving the antibiotic effectiveness of currently available drugs and reaching a sustainable antibiotic use is in the public interest. This could adversely lower the investment in research and development needed to identify new techniques and resources for the future. In this chapter, I argue that the externalities involved in using antibiotics, as described above (as well as antivirals and possibly vaccines), warrants public intervention, now as well as in the future, to correct the incentives of economic agents and to secure support for the research and development of new resources and techniques that will be needed in the future.

I concentrate mainly on the use of antibiotics and begin by presenting bioeconomic models that are used in the literature to address the socially optimal (normative) and market (positive) aspects of antibiotic use.[1] Thereafter I turn to the economic agents, providers, and users of antibiotics and identify possible strategic interactions among agents, and discuss how this may affect antibiotic resistance, in particular, and public health, in general. Open research questions are highlighted that merit future attention from the scientific community.

Bioeconomic Modeling: Normative and Positive Approaches

The network of economic agents involved in the provision and consumption of antibiotics and vaccines is relatively complex. It is particularly intricate in relation to the antibiotic consumption of an individual, which may occur

---

[1] Use of vaccines and antivirals implies similar positive and negative externalities as those present for antibiotic use.

throughout any given year as well as over an individual's entire life span, as opposed to vaccination, which occurs at specific periods of time (e.g., before the flu season, at a particular age). Figure 8.1 illustrates the linkages between economic agents as well as the potential feedback effects from epidemiology related to antibiotic production, provision, and consumption. Economic agents (e.g., patients, physicians, pharmaceutical companies) are embedded in the epidemiological environment where the transmission of infection occurs, and which comprises various resource pools of antibiotic treatment effectiveness. Antibiotic use in animals may also affect this environment. The network is similar for antivirals and vaccines; however, hospitals play a lesser role.

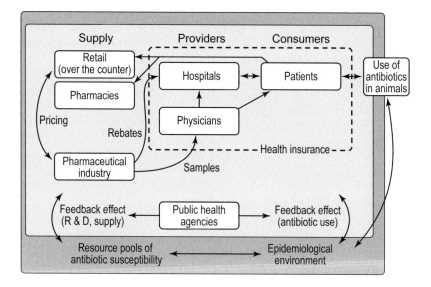

Figure 8.1 Schematic of interactions that occur in the epidemiological environment as a result of disease transmission, involving diverse economics agents. Anti-infective drugs can be characterized by treatment effectiveness, which vary as a function of antimicrobial use and in response to exogenous factors (e.g., microbial mutation). Eight groups of economic agents can be distinguished: on the supply side, the pharmaceutical industry sets the pricing strategy for anti-infective drugs, which impacts pharmacies and retail sales. On the demand side, patients, as the end users, interact with physicians in out- and inpatient (hospital) settings. This interaction is, in turn, affected by the coverage offered by health insurance companies. When necessary, antimicrobial drugs are prescribed by physicians. Public health agencies may regulate antimicrobial drug use. A complex relationship thus determines antimicrobial use, characterized by interactions between these diverse economic agents, the resultant pricing strategies, and availability (e.g., free samples, rebates, existing stock in hospital settings). This relationship impacts disease transmission and treatment effectiveness, which in turn affects both supply (research and development of new drugs) and demand (willingness to pay). As a final note, meat consumption of animals treated with antibiotics can also feedback into the human epidemiological environment.

To address the positive and normative aspects of antibiotic use and identify the externalities involved, bioeconomic research has relied on simplified epidemiological models (for an exhaustive review on the economic literature of biological resistance, including insect resistance to genetically modified crops, see Laxminarayan and Herrmann 2015). Two different classes of biological models can be distinguished in the bioeconomic literature:

1. Deterministic compartment models of disease transmission, where individuals can be susceptible to infection (S), infected (I), or resistant to infection (R), as in the SIS or SIR model (e.g., Wilen and Msangi 2003)
2. Probabilistic models of antibiotic resistance (e.g., Laxminarayan and Weitzman 2002)

In deterministic models, antibiotic resistance is driven by natural selection and the emergence of "*de novo*" resistance, which is proportional to antibiotic use (Bonhoeffer et al. 1997). In probabilistic models, resistance may occur due to the mutation of bacteria. Clearly, all these mechanisms can drive antibiotic resistance, rendering the quality of any normative or positive bioeconomic analysis contingent on the specificity and representativeness of the biological model used.[2]

The seminal paper by Laxminarayan and Brown (2001) analyzes the optimal use of two antibiotics to fight the propagation of infection over time, when the economic objective is to minimize the intertemporal social cost of infection. Each antibiotic is assumed to have its own resource pool of antibiotic effectiveness.[3] When the cost of infection and treatment cost are not an issue, the antibiotic with the highest level of antibiotic effectiveness should be used first until that point in time when both antibiotics have the same level of effectiveness. As of this point, antibiotic use is such that the levels of antibiotic effectiveness remain identical for both. When there is a higher treatment cost of infection with a particular antibiotic, this antibiotic should not be used initially. Optimal use involves using only the cheaper antibiotic at first, lowering consequently its level of antibiotic effectiveness. It is only when the opportunity cost of using only the cheaper antibiotic becomes too important that both antibiotics should be used jointly. Because of the economic intertemporal optimization, these results differ from those obtained in the epidemiological scenarios of Bonhoeffer et al. (1997), who found joint antibiotic use preferable to using one sole antibiotic during an initial phase.

---

[2] Antibiotic susceptibility, the mirror image of antibiotic resistance, is generally defined as the proportion of infected individuals in the overall infected population that are infected with the drug-susceptible strain. An alternative, but not equivalent, measure of antibiotic resistance is related to the number of antibiotic doses necessary to cure an infection (Howard 2004).

[3] The model is derived under the simplifying assumption that there is no cross resistance, nor multidrug resistance, as well as a zero fitness cost.

Wilen and Msangi (2003) present a bioeconomic model in which only one antibiotic is available to fight a given infection. However, here, antibiotic effectiveness is considered a renewable resource: resistant bacteria suffer a fitness cost because they die at a higher rate than susceptible bacteria when no antibiotic treatment is applied. Hence, a low enough antibiotic use allows antibiotic effectiveness to recover.[4] Wilen and Msangi show that the optimal path is characterized by treating the overall infected population during an initial period, followed by a singular, interior control, where only a fraction of the infected population gets treated. Such a particular treatment rate tends to balance the selective pressure exactly due to antibiotic use and the fitness cost of resistance.

Herrmann and Gaudet (2009) extended the model by Wilen and Msangi (2003) and, in addition to the normative analysis, conducted a positive analysis in which antibiotic producers have open access to a renewable, common-pool resource. This market structure represents a benchmark for a generic industry in which no patent protects access to the resource pool. Hence, no producer has an incentive to preserve antibiotic effectiveness by lowering its sales, as this benefit would have to be shared with all the other producers, and thus is another example of the tragedy of the commons. In their model, Herrmann and Gaudet (2009) show that depending on the production cost, which at equilibrium equals the price of the antibiotic, antibiotic effectiveness may be higher or lower than what would be socially optimal in the steady state. The intuition behind this result is that a high price for the antibiotic lowers the quantity in demand. This benefits the maintenance of antibiotic effectiveness, but it does not account for the positive externality which results from antibiotic use: fighting the infection.

In another scenario, Herrmann (2010) considered what happens when a monopolist sells an antibiotic, while a patent is pending, to protect market access. Under this type of market structure, no externality or free-riding problem in relation to the pool of antibiotic effectiveness exists, at least as long as the resource pool associated with the patented antibiotic is not connected to any other antibiotic. In contrast to the social optimum, however, antibiotic effectiveness as well as infection are desirable resources, from the monopolist's point of view, because they represent the quality and potential market size of the drug, respectively. This study shows that the monopolist may want to preserve the antibiotic effectiveness and infection at higher levels. Toward the end of the patent, the monopolist becomes more myopic and prices the antibiotic by giving less weight to the intertemporal aspects of antibiotic effectiveness and infection. The divergence between viewing infection as a valuable resource from the firm's point of view versus undesirable from a social point of view

---

[4] Whether the fitness cost of resistance is positive, zero, or even negative is an empirical question (for a review, see Herrmann and Laxminarayan 2010). A recent publication shows that the fitness cost can indeed be negative (Roux et al. 2015).

becomes apparent when the possibility of patent extensions are considered in response to the problem of antibiotic resistance. Herrmann (2010) shows that a longer patent extension is socially desirable when the prevalence of infection is less an issue or antibiotic resistance is high.

In a more stylized biological model, Mechoulan (2007) also identified the divergence between a monopolist and the "social planner" by showing that the monopolist does not find it profitable to eradicate the disease even though it may be socially desirable to do so. In a highly stylized model that ignores the evolution of infection, Tisdell (1992) showed that a monopolist may indeed correct for the problem of antibiotic resistance, as the monopolist tends to fix higher prices, which lowers antibiotic use, and the monopolist causes less antibiotic resistance. This is a well-known result in public or environmental economics: one distortion (here, the market distortion due to monopoly) may compensate exactly for another distortion (here, antibiotic effectiveness, which is a public good). Clearly, the stylized character of the epidemiological model has an impact on the results. In addition, not all models allow for the same epidemiological events (e.g., disease eradication), which makes the results obtained in the literature difficult to compare.

Considering this caveat on the precise epidemiological context, Laxminarayan and Smith (2006) obtained the general result that the cycling of two antibiotics to treat a given infection (i.e., switching from one antibiotic to another) is only optimal when two conditions are fulfilled: (a) fixed costs are present (e.g., to keep the antibiotic stocked in hospitals) and (b) switching costs from one antibiotic to another exist. This makes the antibiotic treatment cost function nonconvex. When there are no switching costs, a chattering control by instantly switching between one antibiotic and another would be optimal.

In another stylized model that focused on the rise of antibiotic resistance (and abstracted from the prevalence of infection), Laxminarayan and Weitzman (2002) analyzed the benefit of antibiotic treatment heterogeneity. The loss of antibiotic treatment susceptibility of an antibiotic is probabilistic (modeled as a Poisson process) and antibiotic treatment costs differ across antibiotics. Laxminarayan and Weitzman show that the policy of a first-line treatment, in which an infection is fought using the most cost-effective treatment, does not account for the externality of increasing the risk of resistance when concentrating on a single antibiotic. They show that less cost-effective antibiotics should also be part of a treatment policy, so as to diminish the overall risk of resistance: an optimal treatment policy is characterized by the additional treatment cost balancing exactly the benefit of reduced antibiotic resistance of the more cost-effective drug. This economic analysis is interesting because it is the first to capture the rise of resistance due to mutation. It hinges on the caveat that it remains static, in the sense that the optimal mix of antibiotics is not adjusted over time in response to a possibly changing environment. It also abstracts from the positive externality of antibiotic consumption related to the prevalence on infection.

In light of their review (Laxminarayan and Weitzman 2002), the following research questions merit particular attention in the overlapping research field of economics and epidemiology:

- What is the "correct" biological model? Which level of abstraction is acceptable?
- How are resource pools in epidemiological models connected to one another?
- What are the existing substitute treatments among antibiotics and for antibiotics?

Providers of Antibiotics: The Role of Physicians and Hospitals

In developed countries, physicians generally prescribe antibiotics (and vaccines) to patients in both inpatient (hospital) and outpatient settings. One exception is in Greece, where antibiotics are able to be purchased "over the counter." Retail sales of drugs are more common in developing economies, such as India, and is characteristic of the absence of health insurance coverage for a large proportion of the population.

Incentives to prescribe antibiotics in out- and inpatient settings vary greatly (WHO 2014). Since antibiotic resistance is known to be a major problem in hospitals, hospital physicians need to be aware of antibiotic resistance, and good practice guidelines are necessary to avoid the overprescription of antibiotics. In outpatient settings, by contrast, regulations issued by public health agencies are needed to restrict possible overuse by individual physicians, as a means of countering antibiotic resistance.

Inpatient Settings

The intensive use of antibiotics in hospital settings to prevent infections during medical interventions is considered to be a driving factor behind the development of antibiotic resistance. To counteract this, Laxminarayan et al. (2007) found the effective control of the transmission of hospital-acquired infections to be one possible strategy for lowering antibiotic resistance in hospitals.

Whether a hospital has an incentive to control for hospital-acquired infections, especially antibiotic-resistant infections, depends on whether the additional costs related to the treatment of resistant hospital-acquired infections needs to be covered by the hospital or the patient via health insurance. The economic costs associated with antibiotic-resistant infections (e.g., vancomycin-resistant *Enterococci* or methicillin-resistant *Staphylococcus aureus*) have been well documented and account for higher morbidity, higher mortality, and prolonged hospital stays (Laxminarayan et al. 2007; WHO 2014). Even if hospitals are compensated for prolonged hospital stays, hospital-acquired infections lower the productivity of hospitals, thus lessening their ability to focus on first-line medical interventions for which patients are admitted.

Following arguments put forth by Laxminarayan et al. (2007), hospitals may choose to free ride on the efforts of other hospitals that actively engage in fighting antibiotic-resistant infections. This may be especially relevant in patient populations that fluctuate between hospitals (e.g., when a patient becomes infected in one facility and then visits other hospitals), thus creating a spillover effect. To control for hospital-acquired infections and lower their prevalence, direct incentives may be necessary to ensure quality in health care. For example, legislation could require hospitals to provide the public with information on how they manage antibiotic resistance.

Outpatient Settings

The prescription of antibiotics by individual physicians is generally done without testing for the infection's susceptibility, because such tests are costly and time consuming. Furthermore, patients tend to prefer an immediate prescription to satisfy the expectation for an earlier recovery. As a consequence, the prescription rate of physicians may be determined in a matching equilibrium between patients and physicians: if a physician's prescription rate were to be lower than expected, patients would turn to other physicians that prescribe at a higher rate.[5]

Howard (2004) showed that as the level of antibiotic resistance increases, physicians tend to prescribe newer, more expensive antibiotic drugs as first-line treatment. Such a resistance-induced substitution can imply important welfare costs as older, yet still effective drugs could have been used. The increasing use of broad-spectrum antibiotics as a response for high resistance levels to other antibiotics may counterbalance, in welfare terms, the beneficial effect of decreasing the overall use of antibiotics (Howard 2005). Clearly, prescription decisions taken by physicians also depend on their remuneration scheme as well as on health insurance coverage. Masiero et al. (2010) find empirical evidence that a fee-for-service and a salary remuneration are related to higher levels of antibiotic use compared to capitation compensation.

This analysis highlights the need to address the following research questions:

- What are the incentives and disincentives of hospitals to control for hospital-acquired infections?
- What kind of coordination is necessary between hospitals to optimize efforts to encourage cooperation in fighting regional prevalence of antibiotic resistant infections?

---

[5] Such a matching equilibrium has been described by J. Albert in a 2015 working paper, "Strategic Dynamics of Antibiotic Use and the Evolution of Antibiotic-Resistant Infections," where Albert determines the optimal number of providers (physicians) by accounting for arbitrage between temporarily lower prevalence of infection and higher antibiotic resistance in the future.

- Can public information on the performance of hospitals, in terms of managing antibiotic resistance, sustain the right sorting of patients?
- Do economic agents utilize a sufficiently long planning horizon to account for cost-efficiency correctly?
- How does the retribution scheme of physicians affect the prescription of antibiotics?
- Is there any evidence that hospitals vary widely in hospital-acquired infections?

Patients' Demand for Antibiotics

Individuals suffering from bacterial infections are the ultimate end users of antibiotics. Their access to the resource pool of antibiotic effectiveness may, however, be limited by the practice of mandatory prescriptions and the access to and coverage of medical treatments (cost sharing) by health insurance plans.

From an individual patient's point of view, the effectiveness of antibiotic treatment is exogenous, and has been modeled in the literature as a quality aspect of the drug (Howard 2004; Herrmann 2010). When prescribed by a physician, an antibiotic treatment generally represents the optimal strategy for the individual patient, as the patient does not account for the negative externality of possible antibiotic resistance in the future.[6] As outlined above, overall antibiotic consumption creates a selective pressure for resistant bacteria. This selective pressure may be enhanced if patients do not comply with the antibiotic regimen or if resistance arises *de novo* (Bonhoeffer et al. 1997).

Given that antibiotic effectiveness represents a quality aspect of the antibiotic and may affect antibiotic demand positively, a feedback effect for society is created on the demand side that is beneficial: it may slow down the use of antibiotics, even in the case of open access occurring to the resource of antibiotic effectiveness. Indeed, the patients' willingness-to-pay for an antibiotic should be a function of its effectiveness. When antibiotic effectiveness decreases, so does the willingness-to-pay, which diminishes the antibiotic use, and hence slows down the decrease in antibiotic effectiveness. This pattern drove the results in Herrmann and Gaudet (2009) and allows convergence to a strictly positive steady-state level of antibiotic effectiveness (which may nonetheless be suboptimal), thus avoiding the complete exhaustion of antibiotic susceptibility. Such a feedback effect is akin to cases of more classical resources (e.g., halieutic ones). For instance, when consumers understand issues related to the well-being of dolphins, they are likely to reduce their consumption of tuna to reduce the risk imposed by tuna fisheries on the safety of dolphins. A similar effect relates to the consumption of fossil fuels and the consumers' willingness

---

[6] The decision to consult a physician, and purchase and take the antibiotic stems from a dynamic optimization problem. Indeed, an individual first has to decide whether to consult a physician or wait for the infection to clear naturally, which may depend on the morbidity cost.

to pay extra for less-polluting fuels (e.g., biofuels). To which extent such quality consciousness enhances a more sustainable resource use is, however, an empirical question.

One way to reduce the demand for antibiotics resides in the avoidance of infection in the first place. Private decisions on social distancing on an individual level have been analyzed by Fenichel (2013) in a disease compartment model, where immunity to disease arises once an infected individual has recovered (i.e., in the case of viral infections). Following Fenichel (2013), public policies that demand ad-hoc distancing of infected individuals, but which do not account for the health status of individuals, may lead to welfare outcomes that are worse than decentralized outcomes in which each individual evaluates his/her own risk of becoming infected. This is because preventing immune individuals from coming into contact with others increases the likelihood of healthy individuals coming into contact with infected individuals.

Open research questions related to the demand of antibiotics include:

- How does awareness or information about antibiotic resistance affect outpatient demand?
- Is the patient able to wait so that antibiotic demand can be derived in a dynamic context, where the patient accounts for future levels of antibiotic effectiveness and disease prevalence?

Pharmaceutical Industry and the Innovation of Anti-Infective Drugs

In the past, the innovation of antibiotic drugs has followed the pace of increasing antibiotic resistance. Antibiotic resistance can even be considered as having spurred the innovation of new drugs: since the development of penicillin in 1928, 14 new classes of antibiotics have been developed. However, following Coates et al. (2011), the development of completely new antibiotic classes has failed in recent years because of their toxicity to humans. Becker et al. (2006) suggest that no new classes can be developed in the future, as all "broad-spectrum antibacterials have been discovered" (Becker et al. 2006:191). More recent research points to possible new avenues to combat the rise of antibiotic resistance of gram-negative bacteria. Future drugs may attack the protective barrier of bacteria, instead of the bacteria itself (Dong et al. 2014). However, before these drugs enter the market, time-consuming tests for drug approval must still be undertaken.

Given the difficult setting for the development of new antibiotic classes, antibiotic producers concentrate on the development of antibiotic analogues (which belong to existing antibiotic classes). Whether antibiotic analogues are necessarily linked to the same resource pool or whether antibiotics that belong to different antibiotic classes may be connected to a common pool remains an open epidemiological issue. Laxminarayan et al. (2007) propose grouping antibiotics in "functional resistance groups" because of possible inconsistencies that may occur between antibiotic classes and resource pools of effectiveness. Clearly, incentives to innovate a new drug depend on the following:

1. Innovation costs need to be lower for antibiotic analogues, compared to antibiotic classes.
2. How a new drug will be possibly connected to a common resource pool shared with other antibiotics.
3. Which existing treatment substitutes are already available to limit the monopoly power due to patent protection.

Herrmann et al. (2013) analyzed in a stylized context how a pharmaceutical firm decides to distance its drug from a common pool of antibiotic effectiveness when no substitute treatment to the new drug is available. Here, markets are completely separated on the demand side, like antibiotic markets for use in animals and humans, whereas on the supply side they are only possibly connected.[7] While the patent holder of the new drug serves only one market, the other is served by a generic industry. Herrmann et al. find that a firm's incentive to incur a higher innovation cost is determined by the marginal impact that is avoided by the generic industry on the common resource pool of antibiotic effectiveness. Furthermore, the lower the distance of the new drug to the common resource pool of antibiotic effectiveness, the heavier the monopolist discounts its future profits.

The particular nature of antibiotics (i.e., linked on the supply side via common pools of antibiotic effectiveness and on the demand side via substitute treatments) creates potentially nontrivial strategic interactions between pharmaceutical producers. Indeed, an imperfectly competitive environment characterizes the pharmaceutical industry, such that each producer should account, at least to some extent, for the investment and pricing decisions related to antibiotics undertaken by its competitors.

As a result, as the developer of a new antibiotic drug is likely not the only claimant of the resource pool of antibiotic effectiveness to which the antibiotic is related, the question arises as to the optimal patent breadth of antibiotics.[8] Extending the breadth of antibiotic patents may collude with common antitrust laws, as mentioned by Laxminarayan et al. (2007). However, increasing the breadth of patents could increase welfare, because a sole claimant to the resource pool should care more for its sustainable management. Laxminarayan et al. (2007) go even a step further by claiming that *sui generis* rights (i.e., rights

---

[7] For an example of a possible link between the supply side (via a common pool) of antibiotic effectiveness and separated markets on the demand side, consider the case of the U.S. Food and Drug Administration, which withdrew in 2005 an antibiotic belonging to the quinolone class for use in poultry water to prevent the spread of fluoroquinolone-resistant infections in humans.

[8] The patent length is also an issue, as the clinical testing of antibiotics is time consuming.

"of its own kind") over antibiotics could be attributed to antibiotic producers once patents have ended, in order to establish an ultimate claimant of the resource pool of antibiotic effectiveness.

Open research questions related to the pharmaceutical industry's incentive to innovate new anti-infective drugs include:

- What is the optimal patent length and breadth for new anti-infective drugs?
- How can we define functional groups among anti-infective drugs that are linked to common resource pools?
- How are new anti-infective drugs connected on the supply side, and how does this influence the incentives for innovation of new drugs?

Conclusion

If we are to preserve the achievements that have been made in public health, we need to combine our understanding of the private incentives of economic agents, which provide and use antibiotics and vaccines, with knowledge about ecology and evolutionary biology. In the past, the development of new antibiotic classes and analogues has been spurred by the evolution of antibiotic resistance. However, the pace of research and development has slowed down over recent years, making the resource pools of antibiotic effectiveness finite to some degree. Public intervention is necessary to correct private incentives that are using up this valuable resource, but this will not be easy to achieve as resource pools of antibiotic effectiveness cannot be controlled independently at local or national levels. Coordination, on both regional and national levels, is thus necessary and, for this, a global approach will be required. Such an approach has only been analyzed in a stylized macroeconomic model, where each country disposes of one resource pool of antibiotic effectiveness to affect, in turn, positively the health capital of its labor force (Rudholm 2002). Similar issues can be foreseen, related to the coordination among countries, as society confronts new, potentially fatal viral infections that can spread easily by modern means of transportation.

9

Interventions to Control Damage from Infectious Disease

Integrating Ecological, Evolutionary, and Economic Perspectives

Paul W. Ewald, Markus Herrmann, Frédéric Thomas,
Sam P. Brown, Philipp Heeb, Arnon Lotem,
and Benjamin Roche

Abstract

Human infectious disease results from many factors (e.g., human behavior, disease organisms, institutions) that often interact as opposing agents in accordance to the investor–exploiter dichotomy. Directing interventions to influence these opposing roles may improve human health by differentially influencing the success of exploiters and investors. Alterations made on one level may change outcomes on other levels and affect the impact of disease on states of health. These interactions need to be incorporated into economic models to inform the assessments of interventions to improve health.

Introduction

Strategies that involve different interests among participants often involve roles that can be cast as investors and exploiters. Investors increase the presence of some resource whereas exploiters gain access to a resource without increasing its presence. These roles can be applied to infectious diseases[1] and the medical activities used to control them. Assessments of these strategies encompass not only the infectious agents and their attributes, but immune function, human behavior, health care institutions, insurance companies, agriculture, the pharmaceutical industry, environmental conditions, and biological evolution. Understanding the interactions among these elements in the context of investors and exploiters may help inform decisions bearing on the control of infectious disease.

The success of interventions to control infectious disease is often quantified by measuring short-term reductions in incidence, prevalence, morbidity, and mortality. Any intervention, however, introduces a change in the environments of disease organisms. Target organisms may evolve in response to these environmental changes, altering variables such as virulence (i.e., harmfulness) and antibiotic resistance. Evaluations of control efforts, therefore, require an integration of ecological and evolutionary effects; influences on incidence, for example, need to be assessed in light of the evolutionary changes in resistance and virulence.

Assessments of alternative investments in disease control depend on the different interests of the affected participating entities as well as on the ways in which biological, social, and economic aspects interact. In a hospital setting, for instance, hospital administrators may benefit from making investments that reduce antibiotic resistance (e.g., restricted use of antibiotics and extra effort to maintain hygienic standards). These investments, however, may come at the cost of care for individual patients and revenue-enhancing activities. If knowledge about the dangers and rates of hospital-acquired infections are made available to the public, the threat of legal action and improvement in reputation of the hospital may make improved hygiene a higher priority. A pharmaceutical company, on the other hand, may benefit if the prevalence of infection is relatively high and resistance to its own antibiotic is low, especially if it is under patent.

From a public health perspective, the optimal outcome may not be associated with elimination of infection. Mild infections may help maintain an effective immune system; in addition, elimination of infection may be both costly and difficult to achieve. Alternative investments in hygiene may provide a more beneficial outcome and allow a substantial prevalence of benign infection, by inhibiting strains with elevated virulence and reducing the need for antibiotic use. Blocking transmission by hospital attendants, for example, may disfavor virulent, antibiotic-resistant hospital strains relative to mild antibiotic-sensitive, community strains, which are brought in by patients upon admission (Ewald 1994). In general, the best outcome for public health policy makers

---

[1] Infectious diseases are defined broadly here to include any disease caused by parasites, which, in turn, are defined broadly to encompass multicellular, unicellular, and subcellular replicating agents that live in or on hosts and cause harm to them.

would be reduced prevalence, virulence, and antibiotic resistance. Categories of interventions that need to be considered in economic models include the introduction of microorganisms (e.g., probiotic bacteria), administration of antimicrobials (i.e., anti-infectives used against microorganisms) or other therapeutic agents, vaccination programs, and environmental changes that influence pathogen transmission. In this chapter we integrate these considerations to illustrate various levels at which the investor–exploiter dichotomy may be relevant to assessments of interventions to control infectious disease.

Microbes as Investors or Exploiters

Parasites can cause disease by using host resources and generating compounds that damage host tissues. In economic terminology, damaging compounds can be considered "goods." The ways in which these goods influence members of the microbial community within a host and between hosts (during pathogen transmission) determines whether the goods are categorized as public, congested, private, or club (see Figure 9.1 and Burton-Chellew et al., this volume). Public goods are available to all parasites in the environment (i.e., they are not "excludable"), and their use does not deplete the use by others in the environment (i.e., there is no "rivalry" for a secreted compound). These concepts can be applied to the population of a parasite species within a host. For example, if consideration is restricted to the within-host environment, the toxin secreted by

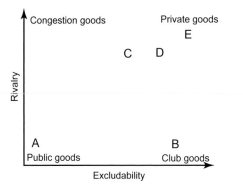

Figure 9.1 Diagrammatic representation of the categorization of microbial goods according to the presence of rivalry (use of good reduces its availability to others) and excludability (benefits of good are not shared). These categorizations can be used to assess expected effects of interventions on pathogens and the diseases they cause. A: Cholera toxin is a public good for the *Vibrio cholerae* within the host. B: Cholera toxin released within a host is not available to the *V. cholerae* in other hosts, so it is a club good when its use is considered among hosts. C: Diphtheria toxin within a host benefits the toxin producers and other *Corynebacterium diphtheriae* in the immediate vicinity and so has an intermediate position in the goods space. D: Diphtheria toxin considered among hosts. E: Compounds that are not secreted, such as those used for cell invasion by *Shigella*, are not shared and hence are private goods.

Vibrio cholerae (the bacterium that causes cholera) is a public good: the toxin released from the bacterium causes cells that line the intestinal tract to generate a diarrheal response, which in turn flushes competing species of bacteria out of the intestinal tract. This effect on competing species allows *V. cholerae* to avoid running through a gauntlet of competitors as they travel through the intestine to the external environment to initiate transmission to other hosts. This transmission benefit is shared with the other *V. cholerae* within the intestinal tract (i.e., the benefit is not excludable), and its use by one *V. cholerae* bacterium does not reduce its use by other *V. cholerae* in the intestinal tract (i.e., there is no rivalry). Consideration of the action of toxin within a host, therefore, leads to its categorization as a public good (location A in Figure 9.1). If, however, the scope of the analysis is broadened to consider the entire cycle of transmission, excludability is present because the transmission benefit is restricted to the group of *V. cholerae* within the host. The toxin is therefore labeled a club good (location B, Figure 9.1). Terminology can be confusing because evolutionary biologists, who consider how levels of selection act on virulence, define goods according to the within-host scale (e.g., Buckling and Brockhurst 2008; Leggett et al. 2014; Zhou et al. 2014). The reason for doing this is that arguments for selection of public goods within hosts must invoke indirect benefits through genetic relatedness of the pathogen population within hosts. Virtually all of the examples of public goods that are discussed in the context of the evolution of virulence, however, correspond to club goods when the analysis is expanded to the population of hosts, because the virulence mechanisms shared by the population of a pathogen within one host (i.e., the "club" of pathogens) are not shared with the pathogens in other hosts.

Other disease-producing compounds lie at intermediate places in the goods space of Figure 9.1. The toxin secreted by *Corynebacterium diphtheriae*, for instance, kills nearby host cells in the respiratory tract, releasing nutrients which can benefit the bacterium that secreted the toxin, but also, to a lesser extent, nearby *C. diphtheriae* (some excludability), and the toxin is consumed in the process (some rivalry). On the basis of within-host effects, the toxin is therefore intermediate between a public good and a private good (location C, Figure 9.1). As is the case with *V. cholerae*, when consideration is expanded to include the *C. diphtheriae* in different hosts, excludability is greater; thus the diphtheria toxin is best considered to have a stronger private goods character (location D, Figure 9.1).

The toxins of *V. cholerae* and *C. diphtheriae* are secreted from these bacteria. Nonsecreted virulence factors are typically private goods. For example, the compounds that *Shigella* bacteria use to invade the intestinal cells are not shared with other bacteria in the intestinal tract (they are excludable); they provide resources that are used solely by the invading bacteria (i.e., they are associated with rivalry) (location E, Figure 9.1). A nonsecreted virulence mechanism could, however, be shareable if it alters the biology of the host systemically in a way that benefits the population of parasites in the host (e.g., an alteration of host behavior that facilitates transmission).

"Public goods" and "private goods" are typically used more broadly in evolutionary ecology than in contemporary economics (e.g., Buckling and Brockhurst 2008; Taylor et al. 2013; Leggett et al. 2014; Zhou et al. 2014). In evolutionary ecology, excludability has been emphasized and variation in rivalry has not been a focus of interest, nor has excludability of effects between hosts. Consequently, the literature of evolutionary ecology generally contrasts public goods with provide goods, with little (if any) reference to congested and club goods. In this chapter we adjust our usage to conform with contemporary usage in economics as illustrated in Figure 9.1.

Interventions to Control Infectious Disease

The health sciences have relied mainly on three categories of interventions to control infectious disease: use of anti-infectives, vaccines, and hygienic improvements. Anti-infectives include antibiotics, used against bacteria; anti-protozoals, used against single-celled eukaryotic parasites; anthelmintics, used against wormy parasites; and, increasingly, antivirals. Vaccines rely on inoculation of parasite molecules that stimulate antibody-mediated or cell-mediated immunity. Vaccines can be prophylactic (i.e., preventive) if administered prior to infection or therapeutic if administered after the onset of infection. Hygienic interventions alter the transmission process. For instance, they can involve disinfecting the skin or environmental surfaces, screening of blood supply, filtration of water or air, or introducing organisms into the host (e.g., probiotic bacteria) that compete with pathogenic organisms or influence immune responses.

Interventions to control infectious diseases alter the selective pressures that act upon causal parasites. In turn, the parasites may evolve to be less controlled by the intervention. This may occur, for example, through increased resistance to anti-infectives or via vaccine escape (i.e., an evolutionary change in the target pathogen characterized by reduced sensitivity to the control by the vaccine).

Evolutionary effects of anti-infectives have been considered most extensively for bacterial resistance to antibiotics for several reasons:

- Antibiotics have been widely used to control bacterial infections for three-quarters of a century, whereas bacteria can evolve noticeably increased resistance within a few years or even a few months after the introduction of treatment.
- Antibiotic resistance is apparent because it diminishes the effectiveness of treatment, which is the focus of medical attention.
- Antibiotic resistance is relatively easy to document *in vitro*.
- Mechanisms of antibiotic resistance are amenable to study.

Most of the attention given to antibiotic resistance within the medical community focuses on evolutionary processes, as the cause of the problem, rather than part of the solution (Chadwick and Goode 1997; Choffnes et al. 2010). The focus, therefore, has been on how to reduce the strength of selection for resistance in bacteria and ways to develop new antibiotics (Choffnes et al. 2010). Guidelines encompass methods to make use of surveillance for resistant organisms, infection control to reduce the use of antibiotics, and prudent use of antibiotics (e.g., by restricting use in agriculture, reliance on narrow spectrum antibiotics, and curtailing inappropriate use of antibiotics) (Stein 2005; Ferri et al. 2015).

The introduction of benign bacteria can alter the community of bacteria in a patient, as is the case with most probiotic treatments. If benign variants of conspecifics (naturally occurring or engineered) are introduced, the intervention lowers the frequency of harmful variants and thus causes at least a short-term evolutionary change. Studies need to be conducted to determine how the persistence of the benign variants can be encouraged. Continual reintroduction or changes in environmental conditions to favor the benign strains are two possibilities. Suppressing harmful organisms through such introductions may be akin to biological control programs in which new species are introduced to control undesirable pest species.

Parasite Goods Involved in Virulence

The effectiveness of an intervention may depend on whether virulence-enhancing compounds are secreted. As suggested above for cholera toxin, expansion of the scope of analysis to consider transmission between hosts in time and space has the general effect of shifting the categorization of the cholera toxin from a public good toward a club good, because fitness benefits of the toxin to other *V. cholerae* are not excludable within hosts but are excludable between hosts. These fitness benefits include the elimination of competing bacteria from the intestinal tract through toxin-induced diarrhea, as mentioned above, and the spread of *V. cholerae* in the external environment, which also results from the diarrhea. Because high toxin production can make a person extremely ill, this latter benefit of high toxin production occurs particularly (a) when water supplies can be contaminated by the fecal material of cholera patients, (b) when sewerage systems are inadequate, and (c) through the movement of the water itself. Within each host, *V. cholerae* variants that produce no toxin should be favored over those that produce high amounts of toxin, because the benefits of toxin production are shared among all vibrios in the intestinal tract (no excludability and no rivalry) and variants that produce little if any toxin are more efficient because they expend fewer resources on toxin production (Baselski et al. 1978, 1979; Sigel et al. 1980; Ewald 1994). In this case, the toxin producers are investors and the toxin-less variants are exploiters. Blockage of waterborne transmission, however, changes these relationships because it changes the relative importance of different transmission routes (i.e., relatively healthy hosts are required for transmission). Where water supplies are protected, lower levels of toxin production are favored evolutionarily (i.e., production of the club good is disfavored), and cholera illness is controlled, even though *V. cholerae* may still be present (Ewald 2002).

Vaccines can provide effective protection by targeting toxins secreted from pathogens. The diphtheria vaccine, for example, controls diphtheria by stimulating antibodies against the diphtheria toxin, which is the most important cause of pathology in diphtheria. The toxin, however, is not produced by all strains of *C. diphtheriae*. If transmitted to a vaccinated host, a toxin-producing strain of *C. diphtheriae* wastes nutrients, making a toxin that is impotent (because the vaccine-induced immune response neutralizes the toxin). Nontoxigenic *C. diphtheriae* do not pay this cost of toxin production. Within vaccinated hosts, they can therefore convert a greater portion of the available nutrients acquired into their own reproduction. The overall consequence is that diphtheria vaccination has led to an evolutionary reduction in virulence of *C. diphtheriae*. The control of diphtheria by this vaccine has thus been extraordinarily successful, not by eliminating *C. diphtheriae* from populations but rather by favoring toxin-less *C. diphtheriae*, which cause little harm to humans while stimulating acquired immunity against both toxigenic and nontoxigenic strains (Ewald 1994, 2002). This vaccination approach, termed the "virulence antigen strategy," illustrates how targeting of virulence antigens can lead to particularly effective control of disease because vaccination favors evolutionary reductions in virulence in addition to reductions in disease incidence.

The cholera and diphtheria examples illustrate how disease can be controlled by a hygienic intervention and by vaccination, respectively, when the virulence mechanism is based on a secreted toxin: a club good in the case of cholera and a toxin that is intermediate between all four categories of goods in the case of diphtheria (see Figure 9.1). Secreted goods may be particularly amenable to evolutionary control because the benefits they provide are often shared within hosts, generating opportunities for less virulent parasites (exploiters) to exploit the benefits of the good that is secreted by the more virulent parasites (investors). Both categories of parasites can therefore be maintained by frequency-dependent selection. If so, the benign strains will already be present in a pathogen population prior to an intervention that favors benign strains. The result can be a rapid evolutionary reduction in virulence, as has occurred in response to diphtheria vaccination. The production of virulence compounds may often be graded rather than categorized as all or none, in which case selection is expected to lead to a graded lowering of virulence rather than a lack of virulence. Evidence indicates that a graded reduction of toxin production occurred in *V. cholerae* in response to protection of water supplies, whereas elimination of toxin production occurred in *C. diphtheriae* in response to diphtheria vaccination (Ewald 2002).

When virulence molecules are private goods, similar evolutionary effects can occur. *Shigella* species, for example, have evolved toward benignity in association with water purification (Ewald 1994). When virulence compounds are not secreted, however, mild strains may be more difficult to maintain by frequency-dependent selection because options for exploiters of the virulence mechanism are more restricted for private goods than they are for shared secreted products. Evolutionary responses to interventions that target virulent variants may therefore be delayed because benign variants (i.e., the exploiters) are not already present at the time of the intervention. When mild variants are present, however, the evolutionary response may be even more rapid than with secreted goods: an immune response to a private good (e.g., a toxin that is physically attached to, rather than secreted from, the pathogen) may directly destroy the virulent pathogen rather than just impose a drain on the investor's fitness (e.g., by neutralizing a toxin).

Evolutionary Stability of Interventions

Medical activities can change the selective pressure acting on target pathogens. Antibiotic treatment, for example, can favor the evolution of antibiotic resistance, and vaccination can lead to the evolution of vaccine escape. The more extensive the activity, the greater the selection pressure for resistance or escape, and hence the greater the loss in the effectiveness of the activity (Figure 9.2). The history of antibiotic use is also the history of resistance to the antibiotics in use (Hede 2014). This linkage is a source of concern because the

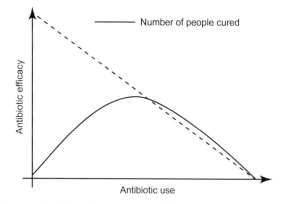

Figure 9.2 Expected relationships between antibiotic use, antibiotic efficacy, and cure rate. As antibiotic use increases, so does the selection for antibiotic resistance. Greater use of antibiotics is thus predicted to be associated with a rise in cure rates that peaks when the positive effect of antibiotic treatment on cure rate equals the negative effect of antibiotic resistance (dashed line). Thereafter higher rates of antibiotic use are associated with decreasing cure rates because of further increases in antibiotic resistance.

generation of new classes of antibiotics has declined as antibiotics have lost effectiveness (Figure 9.3).

The loss of efficacy for antibiotics and vaccines due to evolutionary responses of target pathogens highlights the need to assess whether there are intervention strategies that are more stable evolutionarily. If so, it would be important to specify whether the stability results from limited existing variation in the pathogen population or restrictions on any feasible variation that could arise through genetic changes (as mentioned above for virulence mechanisms that are private goods). New genetic changes hold potential for unforeseen mechanisms that counter an intervention.

Therapies against *Plasmodium falciparum*, the most severe agent of human malaria, provide an illustration. Resistance to antimalarials has increased over the past half century. In response, combinations of antimalarial drugs have been used (The malERA Consultative Group on Vector Control 2011). For this response to be effective, however, early detection is necessary. In countries where malaria is endemic, detection is conducted through a rapid diagnostic test that targets HRP2, aldolase, and LDH proteins of *P. falciparum*. However, recently a mutation on the HRP2 protein that makes the parasite undetectable via the rapid diagnostic test has appeared in Mali (Koita et al. 2012). This association suggests that selective pressures applied by antimalarial combinations could act on avoiding detection rather than selecting for resistance against antimalarials.

Changing the environment to favor benign strains (e.g., by blocking waterborne transmission) is one category of intervention that should stably control disease because it adjusts the action of natural selection to accord with public health interests. Use of vaccines that selectively target virulence-enhancing molecules is another intervention that should be relatively stable evolutionarily,

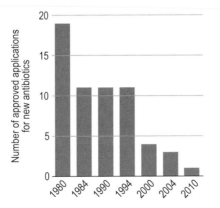

Figure 9.3 Since 1980, the number of drug applications for new antibiotics that have been approved by the U.S. Food and Drug Administration has declined dramatically (after Hede 2014).

because the vaccine introduces a selective pressure that favors mild strains. The target bacterium could evolve an entirely new virulence mechanism, but this evolutionary challenge may be greater than for antibiotic resistance because the resistance genes tend to be present in bacterial populations even before the first commercial use of antibiotics (Pollock 1967; Davies 1994; Courvalin 2010; D'Costa et al. 2011).

The targeting of secreted goods by control efforts may allow for control that is relatively stable evolutionarily, because a less vulnerable secreted good must be generated. During that time, however, intermediate forms of the good may impose a net fitness cost on the pathogen (see Brown, this volume). This hurdle may be responsible for the long-term control of the virulent *C. diphtheriae* by the diphtheria vaccine, which has been in use now for nearly a century.

The targeting of secreted goods may provide a basis for evolutionary stability of new categories of therapeutic interventions to control pathogens. Proteins secreted by pathogens to acquire nutrients, for example, may provide options for therapeutic control of infection. By rendering such molecules ineffective, a therapeutic intervention may compromise pathogens that produce this molecule, relative to those that do not (Brown et al. 2012; see also Brown, this volume). Like the virulence antigen strategy, this virulence interference strategy leaves more benign organisms in the wake of the intervention because the strategy forces the more virulent organisms to waste effort in an ineffective exploitation mechanism. Moreover, this strategy might force the target organism into an evolutionary trap because any variant that increased its investment in the mechanism, as a result of the enhanced need for the resource that the mechanism had previously provided, would be further compromised by this additional ineffective investment (Brown, this volume). Like the virulence antigen strategy, this virulence interference strategy should therefore be relatively stable evolutionarily.

Brown (this volume) supports this hypothesis with evidence from experiments on *Pseudomonas aeruginosa*, in which siderophore molecules that are released from the bacterium to scavenge iron from the environment are targeted by gallium salts. By binding to the siderophores, the gallium interferes with the bacterium's ability to gather iron from the local environment, thereby shifting the competitive balance in favor of less virulent variants that do not waste resources on siderophore production.

This intervention should be relatively stable evolutionarily because siderophores are partially public goods. Administration of gallium eliminates the ability of secretors and nonsecretors to gather iron, but places the siderophore secretors at a selective disadvantage relative to nonsecretors because the secretors incur the costs of siderophore production and secretion (Brown, this volume). Once secretors are eliminated from the population they could, in principle, be regenerated by mutation. In this case, the benefits of their siderophore production would be spread among a high proportion of nonproducers in the population as long as the bacteria are well mixed, and it would therefore be difficult for the producers to increase in frequency in the population. If, however, the bacteria are patchily distributed so that siderophores return iron disproportionately to producers, the goods become less public (i.e., more excludable) and the possibility that the population could be reinvaded with siderophore producers would increase.

One concern with this strategy is that in the vastly larger populations of target organisms in nature, variants which encode a siderophore that binds more selectively to iron than to gallium would be present or could be generated by mutation. If so, the more virulent variants that produce these more selective siderophores would be favored by natural selection in response to gallium use. Increased virulence would then return. Similarly, alternative mechanisms of iron scavenging that were not vulnerable to the gallium treatment could be evolutionarily favored. Overall, as is the case with the virulence antigen strategy, the targeting of siderophores should prove to be more stable evolutionarily than attempts to control bacteria with antibiotics.

Social Context of Interventions

Antibiotic Resistance and Vaccine Escape

Underlying the problem of antimicrobial resistance are conflicts of interests among individuals involved in various aspects of antimicrobial use. Here, too, the components of the system can be cast in terms of goods, investors, and exploiters. Patients benefit from appropriately administered antibiotics, but the consequent evolution of antibiotic resistance poses a cost for patients as they may need antibiotics in the future—a special case of the "tragedy of the commons" (Hardin 1968; Baquero and Campos 2003). Such a conflict of interest creates divergence between goals of health care providers (who focus on preventing or ameliorating disease in patients) and policy makers (who develop guidelines to lessen the development of antibiotic resistance at the population level).

Appendix 9.1 illustrates the criteria used to analyze the degree of vaccine coverage needed to protect a population. The different interests of pharmaceutical companies, insurance companies, agriculture, and health care administrators add to the formidable challenge of developing guidelines and regulations for the appropriate use of antibiotics. One resolution is to create a zone of compromise between too little disease control and too much antimicrobial resistance (Figure 9.4). Such resolutions, however, open the door to influences associated with imbalances of power. In addition, strong incentives are lacking to motivate the pharmaceutical industry to develop new vaccines because of low profit margins: a vaccine is generally used only once by each consumer, and associated costs of development and liability can be high. One approach to this problem involves assessing the potential of target organisms for evolutionary reductions in vulnerability to the intervention and adjusting control

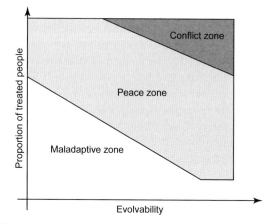

Figure 9.4   The impact of evolvability of antibiotic resistance and treatment consumption on potential conflict between health goals (i.e., reducing illness in a population) and economic goals (i.e., the profit generated from treatment). Three areas can be envisioned: In the *maladaptive zone*, neither goal is met because pathogen transmission does not decrease enough to protect the population and few treatments are sold. In the *peace zone,* both goals are met, at least partially, because pathogen transmission is significantly impacted and numerous treatments are sold. In the *conflict zone,* the goals deeply conflict, because selling more treatments can favor evolutionary changes that negate the value of the treatment.

measures accordingly (Figure 9.4). This would be associated with loss of control or, at best, a continual arms race for a substantial portion of human infectious diseases. Reducing the usage of antibiotics, for example, risks a lack of curative treatment.

These conflicts of interest can be analyzed by considering the problem in terms of goods, investors, and exploiters. Recipients of vaccines are generally considered to be investors in the "good" of vaccine efficacy because their usage generates herd immunity in the population, which in turn inhibits the future transmission of the target pathogen (see Appendix 9.1). Recipients of antibiotics are generally considered to be exploiters of the good of antibiotic therapy because antibiotic usage contributes to antibiotic resistance in the target population. The difference in characterization of the recipients of vaccines and antibiotics, however, results from the tendency for antibiotic resistance to evolve more readily than vaccine escape (i.e., an evolutionary change in target organisms so that they are less controlled by the vaccine).

When vaccine escape can evolve, the categorization of investors and exploiters becomes complicated because the vaccine users are contributing to vaccine escape. In this situation, their designation as investors or exploiters depends on whether the overall concern is generation of herd immunity or vaccine escape. With regard to vaccine escape, those who abstain from vaccination are the investors in the good of long-term vaccine efficacy, and those who become vaccinated are exploiters of this good.

Vaccine escape, however, can positively or negatively affect the overall health of the population depending on the virulence of the escape variants relative to the population of pathogens. When the virulence antigen strategy is used, vaccine recipients are investors regardless of whether one focuses on herd immunity or vaccine escape; vaccine recipients invest in the "good" of disease control generated by both vaccine-induced herd immunity and the shift to benign variants, which is generated by vaccine escape and provides additional acquired immunity for the population. In contrast, if the vaccines are developed from benign variants in the pathogen population or compounds that are more often present on benign variants, the vaccination may selectively suppress the benign variants, and the pathogen population as a whole is expected to escape to higher virulence. In this context, the vaccine recipients are exploiters of the "good" of long-term vaccine efficacy because they benefit from protection while enhancing the generation of damaging escape variants. If the vaccines do not differentially inhibit variants according to their virulence, the effects of the vaccine on the virulence of escape mutants are neutral according to these considerations. The vaccine users, however, are still fostering the spread of escape variants, and in that context are considered exploiters.

Some researchers have argued that the vaccine-induced herd immunity will favor more aggressive strains (Read and MacKinnon 2008). In a laboratory experiment, the use of a vaccine against Marek disease, which reduced mortality but did not block transmission, led to evolutionary increases in virulence (Read et al. 2015). These assessments did not evaluate whether the benign vaccine strains that were used in the vaccine disproportionately suppressed the mild variants in the pathogen population. Still, the hypothesis that incomplete vaccine-induced herd immunity can favor increased virulence needs to be considered as another possible influence of a long-term cost to the population of hosts arising from a short-term benefit to individuals (i.e., the tragedy of the commons applied to vaccination.) In this case, as in the case of vaccines that selectively inhibit benign strains, vaccine users would be investors in the good of herd-induced protection but exploiters in the sense that they are fostering the spread of more virulent escape mutants.

Similar complexities arise when the investor–exploiter concept is applied to antibiotic resistance. The general tendency is to consider antibiotic users as exploiters who contribute to antibiotic resistance. In this context, individuals who forgo use of an antibiotic can be considered investors in the "good" of antibiotic efficacy, because their abstinence from antibiotic treatment contributes to the usefulness of the antibiotic in the future by retarding the development of antibiotic resistance.

This portrayal emphasizes the problem of antibiotic resistance but oversimplifies the details of the investor–exploiter association. When no antibiotic resistance occurs in a pathogen population, antibiotic users are investors in the good of disease control, because the overall effect of antibiotic use is to reduce the prevalence of the infecting organisms, and thereby protect treated individuals and those who would have been infected without treatment. Antibiotics will also tend to act most strongly against the more virulent variants in the pathogen population, because they are most likely to cause disease that is sufficiently severe to motivate infected individuals to obtain antibiotic treatment. When pathogens are susceptible, antibiotic treatment should favor evolutionary reductions in virulence (Ewald 1994). As antibiotic resistance is present, antibiotic users act less as investors and increasingly as exploiters because their suppressive effect on transmission and virulence decreases and their exacerbating effect on antibiotic resistance increases.

Profitability for Industry

Interference therapies, such as siderophore sequestration, would be economically favorable for research, development, and marketing because they would require repeated application. This very same aspect, however, is less attractive from the patient's point of view, because such therapies would be more costly and inconvenient than interventions that involve less persistent use. They would also likely require joint use with other therapies, further reducing their attractiveness for patients relative to therapies that involve just one drug.

In contrast, virulence antigen vaccines would need to be administered much less often, generally once or twice during a person's lifetime. This could reduce the economic incentive for research and development, but would make this intervention attractive to patients. Similarly, from the perspective of public health investments, virulence antigen vaccines promise a large health return on the investment in vaccine development and administration. The diphtheria vaccine, for instance, has been more successful in this regard than any other vaccine, with the exception of the one vaccine that globally eradicated its target pathogen: the smallpox vaccine.

Integrating Biological and Epidemiological Considerations into Economic Models

Whereas bioeconomic studies have focused on the economic objective of the agent (e.g., patients, pharmaceutical companies), purely epidemiological studies concentrate primarily on treatment protocols of anti-infective drug use and techniques. As a result, epidemiological studies may target the prevalence of infection and effectiveness of treatments at a future point of time (e.g., 12 months from now), whereas economic studies analyze the use of drugs by various economic agents (e.g., a patient, physician, hospital manager, or public health agent), or of society as a whole. Furthermore, economic studies determine the welfare that accrues to the agents or society. A further distinction resides in the fact that treatment protocols do not necessarily involve optimization of an agent's or society's objective, whereas economic models generally do.

Economic models assume that an economic objective is pursued by agent $i = \{1, 2, ...\}$ or society $i = 0$ at time t in the context of infection prevalence, $I(t)$, and other epidemiological variables such as treatment effectiveness, $E(t)$. The state variables of the economic objective are $I(t)$ and $E(t)$ (for a discussion of the economic agents that can be involved, see Herrmann, this volume). Given the intertemporal aspects of disease communication and prevalence, the economic objective of an agent may cover a period of time; for instance, a monopolistic firm that sells a new antibiotic benefits from a patent with a limited time horizon. The agent is free to choose from a set of given treatment opportunities for the disease $f(t)$, which we denote the control variable. Notice that control and state variables, $f(t)$, $I(t)$, and $E(t)$ can be vectors, depending on the epidemiological context. For instance, $f(t)$ includes the different treatment techniques (antibiotic use, classical vaccination, virulence antigen vaccination) and $I(t)$ comprises all the infections that are analyzed. We assume that the evolution of infection and other epidemiological variables follow the law of motion:

$$\frac{dI(t)}{dt} = G\left[I(t), E(t), f(t)\right] \tag{9.1}$$

$$\frac{dE(t)}{dt} = H\left[I(t), E(t), f(t)\right], \tag{9.2}$$

where functions $G[I(t), E(t), f(t)]$ and $H[I(t), E(t), f(t)]$ are determined by the epidemiological context (e.g., see Bonhoeffer et al. 1997) as well as the initial conditions, $I(0) = I_0$ and $E(0) = E_0$. Clearly, the alternative treatment strategies and their degree of evolutionary stability discussed above would impact on the functional forms, $G[I(t), E(t), f(t)]$ and $H[I(t), E(t), f(t)]$.

We denote $W_i[(I(t), E(t), f(t)]$ the instantaneous welfare at time t of economic agent i ($i > 0$). An example of this would be the instantaneous profit made by a generic pharmaceutical producer or the social welfare of the overall population (e.g., individuals and firms, $i = 0$).

The objective Γ_i of an economic agent, $i = \{1, 2, ...\}$, or society, $i = 0$, allows us to control for the spread of a disease, $I(t)$, and its virulence (or bacterial resistance), $E(t)$, via treatment opportunities, $f(t)$, and takes the form:

$$\Gamma_i = \max \int_0^{T_i} e^{-r_i t} W_i\left[I(t), E(t), f(t)\right] dt, \tag{9.3}$$

subject to the laws of motion and initial conditions (I_0, E_0) specified above. Notice that $f(t)$, the control variable, should include all treatment opportunities that are available to the particular economic agent or society as a whole. The planning horizon, T_i, and discount rate, r_i, may depend on the economic agent. More myopic agents may use higher discount rates (thus valuing the future less) or shorter planning horizons.

In the objective above, we assume a deterministic context in which the evolution of epidemiological variables can be perfectly foreseen and all treatment opportunities are known. The model could also account for uncertainty related to model parameters (e.g., affecting the speed of the evolution of infection prevalence or treatment effectiveness).

Intergenerational Equity Issues Related to Discounting

The utilitarian objective of aggregating the instantaneous welfare of agent i or social welfare ($i = 0$) gives more weight to the near future and less weight to the long run. This is done through the discount factor $e^{-r_i t}$.

While such a time preference for the short run clearly characterizes the behavior of economic agents, it can be criticized from a social perspective ($i = 0$). Economists refer to such discounting as the dictatorship of the present (Chichilnisky 1996). In our context, it may allow for lower levels of infection prevalence and lead to higher levels of infection in the long run. This clearly raises an issue of *intergenerational equity*.

Discounting future welfare of agents, however, reflects the arbitrage that a dollar invested in health care today could have been better invested elsewhere in the economy (e.g., public education, public transport systems), to increase societal welfare. Furthermore, exponential discounting (as it is done in the utilitarian objective above) is time consistent; that is, the optimal policy, $[(f^*(t)]$, remains unchanged if the optimal state path, $[(I^*(t), (E^*(t)]$, is followed and the objective optimized again.

To address the problem of potential dictatorship of the present, a low (even zero) discount rate can be used from a social perspective. Weitzman (1998) proposes that the far future should be discounted at its lowest possible rate. It has also been suggested that discounting should depend on the state and control variables. Le Kama and Schubert (2007) propose endogenous discounting for intertemporal models that depend on the environmental quality. In particular, a lower environmental quality is associated with lower discount rates, thus giving more weight to the future while keeping the solution time consistent. In our context, the discount rate would be positively related to antibiotic treatment effectiveness, for instance, and could be easily integrated in the above analysis. However, it cannot be forced upon private economic agents.

Time Horizon May Depend on the Type of Infection Analyzed

From an economic perspective, the time horizon, T_i, depends on the economic agent that is analyzed. In the case of a firm selling a new anti-infective drug, the time horizon corresponds to the patent length. Here, the economic objective is to maximize intertemporal profits (i.e., the aggregated profits over the patent length) while the patent is pending. When T_i is finite, however, one should generally account for a bequest function, to capture the sum of profits made after the patent has been suspended and generic firms have become competitors to the incumbent firm.

In some cases, a relevant time horizon is apparent from current knowledge about the problem for a health agency or pharmaceutical company. In the development of the influenza vaccine, for example, assessment of antigens is made annually and the vaccine is altered accordingly. Here the time horizon for the submodel would be one year to allow the investments to span the duration over which each vaccine is to be used, and the intertemporal objective Γ_i would comprise the sum of annual objectives.

New Alternative Treatments May Surface and Others May Become "Extinct"

The arrival of new anti-infective drugs is uncertain as is any research and development outcome. However, the money invested in research and development is dependent on the given epidemiological context. For instance, higher levels of antibiotic resistance may spur efforts made by private firms, but this may be insufficient, as the current antibiotic crisis demonstrates. More profitable opportunities may arise for the industry (e.g., the development of drugs for chronic diseases) or uncertainty in how a newly developed antibiotic might interact with other existing drugs on the market (e.g., common pools of antibiotic treatment effectiveness) may impede work.

The economic methodology can serve as a useful tool to inform public decision makers on the social value of managing existing drug therapies and developing new ones. This analysis can be carried out based on the existing market structure in the pharmaceutical industry. Comparing the outcome of this analysis with what would be socially optimal, it is possible to determine whether public subsidies would be needed to encourage the level of research and development at a high enough level, so as to resolve the anti-infective crisis.

Figure 9.5 illustrates a possible evolution of infection prevalence in relation to antibiotic treatment effectiveness, assuming a typical SIS disease transmission model, when two different objectives are followed. The x-axis measures the prevalence of infection, the y-axis the level of antibiotic effectiveness, and the z-axis the intertemporal welfare related to two bioeconomic objectives: Γ_0 and Γ_1. Let trajectory (0) refer to evolution of infection and antibiotic effectiveness when society pursues the objective of using an existing antibiotic to minimize the social cost of infection, and trajectory (1) be associated with a monopolist selling the antibiotic. The vertical columns refer to the welfare that accrues to the potential consumers of the antibiotic (infected and uninfected individuals) and the producer or providers of the antibiotic (pharmaceutical company, hospital, physician).

In this representation, the prevalence of infection attains lower levels in the social optimum as compared with the trajectory implied by the monopolistic use of the antibiotic. This occurs in typical disease transmission models when

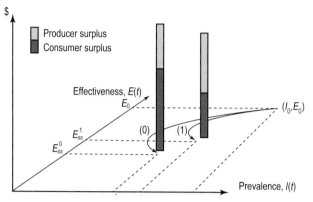

Figure 9.5 Trajectories of infection prevalence and antibiotic effectiveness, leading from initial state (I_0, E_0) toward steady states (I_{ss}^k, E_{ss}^k), with $k = 0, 1$. Aggregated payoffs for consumers and producers are also shown. Trajectory (0) shows the socially optimal evolution ($k = 0$); (1) shows the evolution under monopoly (private optimum, $k = 1$). Note: Trajectories (0) and (1) shown here in state space (I, E) are characterized by an undershooting pattern, which allows infection levels to fall temporarily below their long run steady-state levels. This typically occurs in SIS models, where infections may evolve nonmonotonically, while effectiveness may decrease monotonically (for more details, see Wilen and Msangi 2003; Herrmann and Gaudet 2009).

more antibiotics are used initially in the social optimum. As a consequence, the level of antibiotic effectiveness can also be lower in the short and long term, as compared with the monopoly regime. This can be socially desirable if the social cost of infection sufficiently outweighs the social value of antibiotic effectiveness. The vertical columns show how intertemporal welfare is divided among the population (infected and uninfected "consumers") and the producer (the monopolist). The sum of the producer and consumer surplus is necessarily higher in the social optimum as compared with the monopoly outcome. Notice, however, that the producer surplus is higher, while the consumer surplus is lower in the monopoly outcome as compared with the social optimum.

Such an analysis may appear simplified as it relies on deterministic disease models (Herrmann and Gaudet 2009; Herrmann 2010). However, it does allow us to determine the long-run evolution of variables of interest (e.g., infection prevalence and antibiotic effectiveness), as different economic objectives are followed. When applied to a newly developed drug, it provides information on the potential market size for a drug (infection prevalence) and the quality of a drug (antibiotic effectiveness) that will result when a private economic agent, such as the monopolist selling the drug, maximizes intertemporal, aggregated profits. Thus, this analysis delivers information about the potential profitability of a newly developed drug for a private firm. The private return (or profit) expected from investing in research and development will generally be lower than the social return (social welfare), which points to the potential need to provide incentives to spur the innovation of new drugs.

When the analysis is applied to an existing drug, it can show the critical level at which antibiotic effectiveness is reached, making this drug extinct and necessitating the development of a new drug. Again, the question arises whether the expected private return is sufficient to incentivize the private firm to invest in further research and development. Our analysis here shows that the incentives to develop a new drug cannot be disentangled from its profitability, and hence its awaited market size (infection prevalence) and evolution of quality (effectiveness) of the drug. Public intervention is not only needed to encourage research and development, but also at the level of using the drug once it has been developed.

Integration of Levels

One of the advantages of the sort of model described above is that it can incorporate the spectrum of interactions at play in a complex system, such as human health care, while being flexible with respect to changing conditions, including those attributable to evolution. As discussed, investor–exploiter interactions occur at various levels (Figure 9.6; see also Herrmann, this volume): the outcome of interactions at one level can influence the outcome at others.

In the case of diphtheria, the within-host growth advantage of toxigenic bacteria (toxin investors) over nontoxigenic bacteria (toxin exploiters) favors a high frequency of toxin investors in unvaccinated populations of humans. Coughing can transmit both strains to new hosts, where a new round of growth perpetuates the dominance of toxin investor. The epidemiological record indicates that this perpetuation is stable so long as vaccines are not introduced. In vaccine recipients (i.e., human vaccine investors), however, toxin investors are selectively disadvantaged because they waste resources on the production of ineffective toxins. This leads to eventual stable dominance by benign exploiter strains, which generate additional immunity to all strains because all strains have essentially the same antigenic composition, after toxin-associated

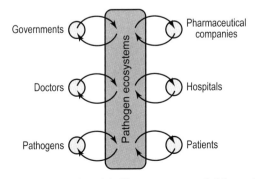

Figure 9.6 Pathogen-mediated social dilemmas are coupled by pathogen ecosystem dynamics (prevalence, resistance, and virulence), wherein agents interact with each other.

antigens are taken into account. In a vaccinated population, the dominance of toxin exploiters is virtually complete, leading to the virtual elimination of diphtheria, even though *C. diphtheriae* are still widely distributed (Ewald 2002). This elimination of symptomatic infection virtually eliminates exposure of the bacteria to antibiotics because asymptomatic individuals do not seek antibiotic treatment. The use of this vaccine, therefore, virtually eliminates the presence of humans who are antibiotic exploiters (because antibiotic usage is extremely low and hence antibiotic resistance is not a pressing problem). The vaccine causes humans to be antibiotic investors in the sense of investing in lower virulence, because antibiotics will tend to suppress the most virulent strains. This outcome occurs because the infrequent use of antibiotics should be associated with antibiotic sensitivity, and antibiotics need only be used for rare infections that are symptomatic due to a high infecting dose or a novel virulence mechanism. Thus, the virulence antigen strategy converts vaccine users into investors in herd immunity and lower virulence and antibiotic users into investors in lower virulence, and eliminates the exploiter role for antibiotic users. The virulence antigen strategy resolves the tragedy of the commons in both vaccination and antibiotic use. The diphtheria vaccination program demonstrates this resolution by generating not only extraordinary success at controlling diphtheria but an association with a general paucity of antibiotic resistance among *C. diphtheriae* strains (von Hunolstein et al. 2002).

If an economic model analyzing social optima places a priority on health promotion, the model should identify this outcome as optimal because it eliminates diphtheria, even if it results in low profits for the manufacturers of antibiotics and only moderate profits for the vaccine manufacturer and health care workers. The extraordinary success of diphtheria vaccination can thus be attributed to an intervention that disfavors bacterial investors of toxin production and favors the human vaccine investors (i.e., those who agree to be vaccinated).

Although this assessment accounts for three different investor–exploiter pairings, other pairings are also present. Table 9.1 provides a listing of some of these investor–exploiter pairings, often present simultaneously in health settings, that are associated with human infectious diseases. Here, we have investigated only a few of the issues raised by considering these pairings. Incorporating these differences of interest into health decisions remains a major challenge for future efforts to control the damage caused by infectious diseases.

Conclusion

Human infectious disease involves many interacting players that can be cast in the context of investor–exploiter dichotomies, and there is a tendency for goods to be shared and consumed among all players. The interests of players may often conflict and change during epidemiological and evolutionary

Table 9.1 Some investor–exploiter pairings that are often present simultaneously in health settings.

Players	Goods	Investor	Exploiter	Goal of Investors
Pathogens	Extracellular molecules, virulence factors	Pathogens that produce the molecules	Nonproducers that benefit from the molecules	Reproduction, survival, transmission
Patients	Herd immunity Antibiotic effectiveness	Vaccine recipients Patients who refuse antibiotics	People who refuse vaccines Patients who take antibiotics	Patient health, survival
Animal husbandry	Herd immunity Antibiotic effectiveness	Vaccine users People who restrict antibiotic use	Nonusers of vaccines People who administer antibiotics to animals	Disease control and agricultural productivity
Pharma firms	Antimicrobial sensitivity	Developers of antibiotics	Imitators	Maximize profit over finite planning horizons
Hospitals	Reduction of nosocomial infections	Enactors of sanitation	Contributors to lax sanitation	Reduce infections and associated costs
Insurance companies	Client policies	Promoters of vaccination	People who refuse vaccination	Maximize profit
Populations, countries	Public health			

timescales. Investor pathogens may secrete compounds that enhance their acquisition of benefits by themselves as well as by exploiter pathogens. Targeting these compounds may yield vaccines or treatments resistant to evolutionary responses that negate their effects (e.g., vaccine escape and antimicrobial resistance). Vaccine recipients are investors in herd immunity, which provides protective benefits. When herd immunity is very high, the negative effects of vaccines may outweigh the beneficial effects to the individuals: those who do not accept vaccines are exploiters of the herd immunity who contribute to the possible reemergence of the target pathogen. Recipients of antibiotics exploit the therapeutic benefits of antibiotics while contributing to antibiotic resistance in the general population. When resistance genes are not present, however, the users of antibiotics are investors in the control of the pathogen in the population. Consideration of these roles can be critical, for example, in hospital environments, where control of infection with antibiotics can protect the patient population but also contribute to the emergence of antibiotic resistance. In addition to these considerations at the level of microbes and public health are economic incentives of physicians, hospitals, and pharmaceutical companies. Each has its own priorities for investment and exploitation, which may differentially influence health-care activities. These interactions need to be incorporated into economic models that recommend multifaceted investments in health care.

Acknowledgments

We thank Michael Kosfeld, Julia Lupp, and Luc-Alain Giraldeau for comments on the manuscript, and Friederike Mengel for her input on categorization of goods in economics.

Appendix 9.1: Public Health Considerations for Vaccine Usage

For infectious diseases with permanent immunity, such as measles, the critical proportion of a population that needs to be vaccinated to eliminate disease can be calculated based on the basic reproductive ratio, R_0, which is quantified as follows:

$$R_0 = \frac{\beta N}{\sigma + \mu}, \tag{9.4}$$

where N is the number of individuals (assumed here to be entirely susceptible or are infected), β is the transmission rate of the pathogen, σ is its recovery rate, and μ is the host life span. This number represents the number of secondary infections caused by one infectious individual within a host population entirely susceptible during the individual's infectious period. A common goal of any public health strategy is to decrease R_0 below 1, in order to drive the pathogen to extinction.

Vaccination strategies rely on a decrease of the number of susceptible individuals (represented here by N). $R_0 < 1$ if $p < 1/R_0$, where p is the proportion of susceptible individuals. Therefore, the minimal proportion of population to vaccinate for expecting eradication is:

$$Pc = 1 - \frac{1}{R_0}. \qquad (9.5)$$

For instance, measles has an estimated R_0 of 17, which means that $1 - 1/17$, or ~94%, of the population need to be vaccinated to create a herd immunity that prevents infectious invasion. Because transmission is dampened as the proportion of susceptible individuals declines, this threshold of vaccination should be sufficient to eliminate the pathogen. This threshold does not account for population heterogeneity or pathogen evolution, and thus must be considered carefully with regard to the relevance of these factors in the targeted pathogen.

Vaccine uptake and pathogen transmission could be strongly linked in a dynamic manner. Indeed, when a sufficient number of people are vaccinated, and effects of herd immunity are prevalent, the perception of the risk of getting vaccinated may become greater than the perceived risk of getting the infection. Such a decrease in the perceived risk of infection can have a dramatic impact on vaccine uptake if individuals act solely out of self-interest.

This dilemma, known as the "vaccination game," has been well studied: epidemiological dynamics is coupled with a theoretical game approach that considers individuals getting vaccinated according to their self-interest (Bauch and Earn 2004). By considering a simple situation where everybody has the same information, it has been demonstrated that increases in perceived risk of vaccination yield a larger decrease of vaccine uptake for the pathogen with high transmission profile (i.e., large R_0), which is difficult to restore after the end of a vaccine scare. Moreover, this theoretical approach shows that for any non-null perceived risk of vaccination, the expected vaccine uptake is always less than the eradication threshold. Although these results must be nuanced according to the strong assumptions of this framework (especially the homogeneous spread of the information), they nevertheless highlight the importance of considering the feedback between epidemiological dynamics and individual decisions based on self-interest.

Strategies to Counter Exploitation

10

Challenges for Market and Institutional Design when Countering Exploitation Strategies

Gigi Foster, Paul Frijters, and Ben Greiner

Abstract

Cooperation within larger groups is often endangered by incentives to free ride. One goal of market and institutional design is to create environments in which socially efficient cooperation can be achieved. The main point in this chapter is that only considering first-order incentives to cooperate within a larger group may not be sufficient, as subcoalitions display reciprocal behavior despite the incentives to renege. Three related complications are discussed: (a) exploitative behavior is often coordinated in subgroup coalitions, (b) natural within-group resistance to exploitation already exists, and (c) the actions of group members can often only be imperfectly monitored. Given these realities, implications of current research for applied market and institutional design are outlined.

Introduction

A well-known problem for any social group is that some individuals may find it more rewarding to try to steal the product of the group's activities, instead of making whatever investments are required to join the group and participate legitimately in the group's productive activities. There are many forms of such exploitation. In the human world, they range from free riding on publicly provided goods (i.e., consuming such goods without having paid for them), to untrustworthy behavior in markets and corruption within organizations, to rogue jihadist groups stealing resources in unstable regions. In the animal kingdom, examples of exploitation abound. From birds who brazenly steal other birds' parenting resources (Davies 2000), to insect fathers that pass genetic codes onto their offspring that greatly increase the offspring's chance of rising in the social hierarchy while avoiding detection by competing fathers (Hughes and Boomsma 2008), to bacteria that cheat their peers (McGinty et al. 2011), the social structure of animal species leaves room for individually advantageous cheating features (for more animal examples, see Dubois et al., this volume). Humans, clearly, are not the only animals to have hit upon cheating as a potentially viable survival strategy.

Market design "helps solve problems that existing marketplaces haven't been able to solve naturally" (Roth 2015:7). That is, economists as market designers have started to take over a role which had previously been largely occupied by entrepreneurs and lawmakers by not only trying to understand the working of markets but also using that understanding to rewrite the rules of markets in order to fix them when they are broken. Applied to the context of exploitative behavior, market design (or organizational/institutional design)[1] with a social welfare objective seeks to make anyone's participation in an institution or market safe from being exploited. The lessons learned from this endeavor may also be able to inform research about cooperation within and between other species. In designing ecological environments, for example, humans may be able to achieve states that allow endangered species to survive.

To research the underlying social dilemma, behavioral and experimental economists have modeled exploitation as a public goods game (where it is socially optimal but not individually rational for each individual to contribute to the public good) and examined the effects of such things as group size, group composition, the size of free-riding incentives, the possibilities for leadership, and other aspects of the group environment on the level of contributions (for literature surveys, see Ledyard 1995; Chaudhuri 2011). This is different from how cooperative investment and exploitation is modeled in evolutionary biology. In the latter, the (co)existence of investors and exploiters is the outcome of an evolutionarily stable equilibrium in a model that assumes all individuals to be selfish with respect to increasing their chances of reproduction. In (proximate) behavioral economics models, such as the one outlined below, cooperation (i.e., nonexploitative investment) is a choice that cannot be rationalized when assuming purely selfish individuals maximizing their short-term utility. However, empirically, some individuals deviate from the behaviors predicted for rational selfish agents, even in the absence of institutional interventions. These deviations may result from the incompleteness of economists' models of individually rational behavior, which arguably exclude some individual traits that evolved due to the longer-run dependence of individual survival and reproductive success on the social group structures in which individuals operate. Nonetheless,

---

[1] In this chapter, we use the terms "market design" and "institutional design" more or less synonymously. While the targets of economic design differ, the approach is the same: creating rules and incentives such that the market designer's objectives (which are here assumed to be aligned with society's objectives), in terms of outcomes, are met.

our focus here is on the short-term, proximate utility model, to highlight the immediate incentives of individual actors and how to deal with them.

In a typical behavioral economics model, the simplest setup of a situation in which agents may benefit from the investments of other agents can be summarized by the following reward equation. The equation describes the utility payoff of an individual, Y_i, based on the agent's own economic choice, x_i (where a higher value denotes a more prosocial choice), and the choices of others, x_{-i}:

$$Y_i = f_i(x_i, x_{-i}). \tag{10.1}$$

A social dilemma emerges when for some individuals

$$\frac{df_i(.,.)}{dx_i} < 0, \text{ and yet } \frac{\sum_j df_j(.,.)}{dx_i} > 0. \tag{10.2}$$

That is, being more prosocial is associated with a loss to the individual but with a net benefit to the whole group (and being less prosocial has the opposite set of effects). The utility function f_i can differ between individuals such that in equilibrium, some individuals may contribute while others free ride.

Market and institutional design can be thought of as the creation of institutions that alter the relationship between individual and group interests (see also Ostrom et al. 1992). If a social welfare-oriented market design is effective, then it aligns the interest of the individual with the interest of the group. To allow for such institutions, we can extend the reward equation with institutional investments α_i and α_{-i} that run from low to high:

$$Y_i = f_i(x_i, x_{-i}, \alpha_i, \alpha_{-i}), \tag{10.3}$$

with $\alpha_i = \alpha_{-i} = 0$ in the case that no institutions exist, resulting in the dilemma situation described above. An effective market or institutional design alters the dilemma game such that it allows for investments α_i that affect the relationship between x_i and Y_i. At some point the total investments $\sum_i \alpha_i > 0$ are large enough such that:

$$\frac{df_i(.,.,\alpha_i,\alpha_{-i})}{dx_i} > 0. \tag{10.4}$$

In this case, it becomes individually rational to contribute to the public good, even for individuals who would free ride if no institution were in place.

The investment α_i may be costly. It could represent a credible commitment to punish, as in the classic public goods games which feature a punishment option (Fehr and Gaechter 2000); in this case, an individual's contribution to punishment represents a contribution to a second-order public good (for discussion, see Yamagishi 1986). If a sufficient number of people plan to punish selfish behavior, and this information is public, then the overall level of punishment expected for not contributing may be sufficient to incentivize efficient contributions to the first-order public good. A large laboratory experimental literature, starting with Fehr and Gaechter (2000), shows that allowing for costly punishment increases cooperative investments in social dilemma situations. Krueger and Mas (2004) and Mas (2008) provide empirical evidence for the relevance of costly punishment outside the laboratory.

The institutional investment α_i may also be the establishment of a reputation system, such as the feedback systems in online markets like eBay. Giving feedback may be costly: it requires taking the time to log in and give honest comments, which if negative may result in retaliation, creating a further cost. If adequately utilized, however, this type of feedback system may be effective in incentivizing buyers and sellers to behave in a trustworthy manner in the online market platform (see also Bolton et al. 2013).

As a general principle, the function of market design is to provide institutions that either affect the relationship between x_i and Y_i directly, or provide an infrastructure such that investments α_i can affect the relationship between x_i and Y_i at a reasonable cost. Several such options might immediately come to mind: allowing communication and increasing visibility, closing loopholes that can be strategically exploited, or providing more effective punishment options.

However, the social world is considerably more complex than the picture painted above. As a result, direct, simple measures that only address first-order dilemma incentives may be ineffective. Merely ensuring that deviations by a single individual are not profitable may be insufficient and, in the worst case, may even create an effect opposite to what was intended.

To illustrate this, we discuss three types of complications:

1. Exploitation is often carried out in *groups* rather than by individuals. This is because individuals intending to exploit others often must cooperate with each other to do so. Effective institutional design will thus need to provide institutions that steer group dynamics toward inhibiting collusion within exploitative subgroups, while at the same time encouraging cooperative investments within the larger group (or "society").

2. Within many social environments, counterforces are already at work to combat social exploitation. In many cases, these counterforces take the form of altruistic punishers, who prosecute exploiters at their own cost and without formal institutional arrangements. Effective institutional design may need to provide support for these altruistic individuals and be careful not to crowd out their motivation and efforts.

3. In real-world environments, information about others' behavior is usually noisy. Such imperfect monitoring may have a significant impact on the effectiveness of punishment, even with small amounts of noise. Effective institutional design needs to take into account the effects of type I errors (i.e., investors being punished as exploiters) and type II

errors (i.e., exploiters not being punished) under conditions of imperfect monitoring.

Below, we address each type of complication, discuss relevant recent research, and highlight challenges that arise. We conclude with a discussion of market and institutional design based on preliminary research results, review implications for designing enhanced cooperative investment among nonhuman species, and propose directions for future efforts.

Collusion within Subgroups

The current economics literature on social dilemma games often assumes there is a dominant group from which individuals can only break away alone. However, during our long human lives we live in many different groups in which we are shaped, produce, fight, share, love, reproduce, shape the next generation, and die. These groups range from the family in which we are born to the organizations for which we work, the causes we support, the countries and sports clubs to which we belong, and the organizations and families we set up ourselves. We function simultaneously in many groups, sometimes switch groups when the opportunity arises, and sometimes initiate new groups and subgroups that further our interests (Frijters and Foster 2013:169–170).

The main behaviors that can hamper the efficiency and effectiveness of groups are then not only the free riding of individuals, but also the free riding of subgroups who do not pull their weight: the exploitative behavior of cohesive privileged groups that comes at the expense of society as a whole. A prime economic example of such antisocial subgroup cooperation, often termed "rent seeking" in economics, is seen in cartels. Companies in cartels cooperate in their choice of prices such that they all make greater profits at the expense of the larger society (for examples of conceptually similar behavior exhibited by nonhuman animals, see the discussion about "ganging up" in Dubois et al., this volume). Other examples within countries are criminal gangs or the regulatory capture of, say, financial watchdogs by financial interest groups that use that capture to evade monitoring by society as a whole. An example of exploitative coalition formation on the international scale is the phenomenon of individual countries (which are subgroups of "the international community") reneging on agreements about global climate change.

One puzzle here is how these groups manage simultaneously to cooperate within their subgroup and selfishly exploit the larger group. That is, the individuals in these groups behave at the same time selfishly (with respect to the larger group) and in a trustworthy fashion (with respect to their fellow coalition members). How can you trust someone who is observed to cheat someone else at the very same time? When it comes to whole countries breaking away from international agreements, the answer is simple: the prime loyalty of individuals is toward their country, not toward an international agreement. Within countries and especially within individual markets, the answer is less clear.

The concept of "group identity" (Akerlof and Kranton 2005) may be helpful in this context. Differences in behavior toward a subgroup and the larger group (or other subgroups) may be sourced in a higher identification with the smaller group which in turn translates into a higher utility derived from adhering to that smaller group's norms, as opposed to universal norms. It may even trigger stronger punishment within the smaller group, enforcing these norms (Goette et al. 2006).

From a market design perspective, the question becomes how to disincentivize cooperation within an exploitative subgroup to increase cooperative investment within the larger group. With respect to cartels, for example, the literature on competition policy already proposes some methods of combating such collusion: closer supervision, preventing within-subgroup communication, whistle-blower provisions, and so on. Essentially, the existing identity with, and loyalty to, the larger group is harnessed and directed toward establishing and implementing monitoring and punishment mechanisms against any would-be smaller group.

The simple mathematical framework outlined above can be used to model the formation of a subgroup within a larger group such that cooperation within the subgroup occurs at the expense of the production and consumption of the larger group. This conceptualization of the problem accommodates many forms of corruption and collusion. Individuals face the option of being prosocial within a clique, but antisocial toward the group as a whole. There is thus a choice c_{ij}, denoting an action taken by i that affects j (where j is the individual or subgroup favored), which is nonetheless antisocial from the point of view of the group as a whole. The individual or subgroup j can then reciprocate this choice by choosing an appropriate c_{ji}. In terms of the effects of these choices on the individual's material rewards Y, we would have the following situation:

$$\frac{dY_i}{dc_{ij}} < 0, \frac{dY_i}{dc_{ij}} + \frac{dY_i}{dc_{ji}} > 0, \frac{dY_j}{dc_{ij}} + \frac{dY_j}{dc_{ji}} > 0, \sum_k \left\{ \frac{dY_k}{dc_{ij}} + \frac{dY_k}{dc_{ji}} \right\} < 0. \quad (10.5)$$

Expressed in words, these four conditions say that the antisocial choice c_{ij} costs the individual (first condition), that the individual nonetheless gains if the other individual or clique j reciprocates (second condition), that the individual or clique j gains from the antisocial choice and its reciprocation (third condition), and that there is still a net cost to the whole set of individuals from the antisocial choices and their reciprocation (fourth condition). The question is: What institutions can be devised to break the "benefit" to cliques (i and j) of being antisocial from the point of view of the group as a whole?

Two recent studies provide some initial insights into the endogenous formation of exploitative subgroups. In an ongoing study conducted by two authors of this chapter (Foster and Greiner), human subjects are invited to an experimental computer laboratory and take part in a game in which they can earn a substantial amount of cash.[2] Participants interact over 100 rounds in fixed groups. In the first stage of each round, subjects vote over who will get to allocate group resources (a monetary amount) as a dictator in the second stage. This setting is extreme in the sense that the game has an "empty core": cooperative game theory predicts that in this game there is no stable coalition of a subgroup of players that would be robust with respect to counteroffers made by other players. (Given small voting costs, noncooperative game theory, in fact, predicts that every member will vote for him/herself.) The experiment involves three possible treatments:

1. There is no communication before the dictator election takes place.
2. There is a stage in which each group member submits a nonbinding distribution proposal (similar to a cheap-talk election campaign).
3. Proposals are submitted and are binding, such that group members vote for a distribution as proposed by one of their members.

Results from this study show very strong and long-lasting coalitions, the size of which depends on the institutional setting. In particular, when there are no campaigns or only nonbinding proposals, in many cases a minimum majority (i.e., an exploitative coalition containing three out of the five subjects) was observed among whom the elected dictator distributed all resources, to the exclusion of others. Nonbinding counterproposals made by the excluded subjects trying to break up these cliques were often unsuccessful. Only when proposals were binding did groups often end up in so-called "grand coalitions," featuring fair distributions of resources among all group members.

In a way, these results mimic the emergence of clans or families within larger societies. The treatment effects suggest that if market design can affect the informational environment, then it also can affect the emergence and stability of exploitative coalitions. With limited communication between group members, only the current resource owner can communicate credibly through an actual allocation decision. Even if the institutional setup allows for cheap-talk communication, the current dictator has the advantage of being able to support his/her own proposals with evidence, while counterproposals from others in the group lack this credibility. However, when individuals can credibly commit to proposals, there is a lower likelihood that an exploitative subgroup grabs all resources and distributes them among themselves, and a higher probability of the emergence of a fair and equal distribution.

Foster and Greiner make another interesting observation in their experiment. In the treatment with nonbinding proposals, two out of three experiment sessions had to be stopped prematurely, at rounds 50 and 60, respectively. This was because some subjects in these experiments became very upset about being, in effect, permanently excluded from the group's resources, even though

---

2   A first version of the research paper on this study is expected to be published by the end of 2016.

they made generous counteroffers. In looking for possibilities to punish the exploitative behavior that victimized them, they found their own way of breaking the rules and starting a revolution: they delayed their decisions—for which a time limit was not enforced in the experiment—such that the experimental session progressed at an increasingly slow pace and eventually had to be prematurely stopped.[3]

The second study, by Murray et al. (2015), provides evidence about how inefficient coalition formation arises. Here, fairly small groups of subjects (four or six) played a game in which whoever was the leader in a round had to choose a partner in that round who would get a large bonus and who would also be the sole producer for the whole group. Those who were not chosen only received a share of this group production, but not the relatively large additional bonus of being the producer. In a subsequent round, the former partner became the leader, or had a high probability of becoming the leader, which endogenously gave rise to long-lasting partnerships that thrived over many rounds, at the expense of the group as a whole. An overall loss in welfare came about from the fact that the productivity of each person varied in each round, meaning that these sticky partnerships prevented optimal allocation—in which the chosen partner would always be the most productive person in that round—from emerging. The experiment featured the type of back-scratching behavior observed inside major institutions and within clans and other larger groups, a major problem in both developed and developing countries (e.g., Murphy et al. 1991).

The experiments of Murray et al. (2015) included the introduction of market design institutions aimed at breaking up inefficient partnerships once they had formed, by either reducing the bonus that the leader would allocate or by randomly breaking up a coalition and putting someone else in charge as leader of that round. It turned out to be very difficult to split up coalitions: the dominant pattern was that former partners who were randomly broken up would simply reestablish their coalition as soon as they were in the position to do so, often leading to subgroup coalition formation by everyone in the experiment: that is, the whole group became divided into a set of (active or dormant) dyadic partnerships. Even when bonuses became low enough that even the coalition members would be better off without being in a coalition, they largely endured, perhaps due to fear that others would otherwise come to dominate the leadership positions.

So far the main implication from this line of research is that some institutions might be advocated to prevent subcoalitions from forming, while other institutions might effectively dismantle them. In particular, juxtaposing people who are unfamiliar to each other appears to help delay coalition formation, and this can be combined with setting low levels of discretion (high levels

---

[3] To us, as experimenters, this was actually bad news, since we lost experimental control (in the sense of defining and controlling the available strategy set). Nonetheless, it is an important observation.

of monitoring), an institutional backdrop that lowers the benefits of coalition formation. Human warfare provides classic examples of attempts that make it hard for soldiers of opposing armies to fraternize or desert through monitoring. In the eighteenth century, for instance, Frederick the Great used very conspicuous uniforms for several of his troops, so that they would be (a) clearly visible to the enemy and (b) easily spotted should they attempt to cooperate or desert (von Clausewitz 1832/1968:415). Breaking up established coalitions, however, appears to be more difficult.

Existing Counterforces to Selfish Behavior

In many dilemma situations, there exists some spontaneous disciplining behavior. Even in the absence of formal institutions, some people not only invest in the joint group interest (e.g., the public good), but also act as disciplining agents toward those who do not invest. This is true even in nonhuman social animals, in whose societies evolutionary biologists have observed both antisocial behavior and punishment of that behavior, with both explained as the result of genetic competition. As Hughes and Boomsma (2008:5152) state: "The nonidentical reproductive interests of group members inevitably result in individual-level selection favoring cheating and the antagonistic coevolution of cheat suppression."

Casual observation and the results of economic experiments show that this cheat suppression can occur even when disciplining actions are costly. In the background, cheat suppressors are often sustained by embedded social norms, like an individual conscience, that support their behavior. As noted by Frijters and Foster (2013:178) with respect to the enforcement power of such individuals within large groups, even a small number of individuals who are willing to mete out punishment to those who are supposed to be punished might be sufficient to dissuade exploitation by others in the group. Hilbe and Sigmund (2010) and dos Santos et al. (2011) show that when past behavior can be observed, and thus a punishment reputation can be built, the existence of (seemingly) altruistic punishers can be an equilibrium outcome in an evolutionary "meta-game" with purely selfish agents.

Various studies on norm enforcement, starting with Fehr and Gaechter (2000), have found evidence for altruistic punishment. In a typical public good experiment with peer punishment possibilities, punishment is costly, both in terms of the direct costs of executing punishment (in the typical experiment, it costs 1 unit to inflict a punishment of 3 units) and in terms of counterpunishment (often people who get punished then punish back, as a reciprocal action or to discourage further punishment). Altruistic punishment behavior is thought to be driven by an emotional response to norm violations that is acted upon even at a cost to oneself. It is more likely to be observed among those who cooperate most themselves (e.g., Fehr and Gaechter 2002).

In the language of our framework, altruistic punishers have a utility function $Y_i = f_i(x_i, x_{-i}, \alpha_i, \alpha_{i-i})$ such that $dY_i/d\alpha_i > 0$, meaning that the costs of contributing to the punishment institution are more than counterbalanced by the value derived from such a contribution (e.g., through a positive internal sense of "doing the right thing"). A challenge for market design is then to ensure that the prosocial incentives of these altruistic punishers are not crowded out by introduced institutions.

There is evidence that norm enforcers exist in other environments. Bolton et al. (2015) study the interaction of reputation systems and conflict resolution systems and provide a case study of eBay's feedback system and the feedback withdrawal process. On eBay's pre-2007 platform design, after each transaction, both the buyer and seller could give positive, neutral, negative, or no feedback on the trading partner. Feedback was immediately published on the platform. The feedback could only be withdrawn if both transaction partners mutually agreed to a withdrawal. Such a withdrawal process was introduced to the platform to facilitate the resolution of a conflict after it had arisen, through the "making good" of initial offenses and subsequent removal of negative feedback. However, the withdrawal option also provided strategic incentives that actually may have escalated the conflict in the first place. In fact, the possibility of mutual feedback withdrawal makes it a dominant strategy to respond to negative feedback with one's own retaliatory negative feedback, independent of the transaction details. This is because the retaliation is a low-cost means of creating negotiating power that can be applied during negotiations for mutual feedback withdrawal, which may then occur without the original offender incurring the (presumably higher) cost of "making good."

Using field and laboratory data, Bolton et al. (2015) showed that the existence of a withdrawal option may, in this way, hamper trustworthiness, and thereby trade efficiency, on eBay's platform—as well as lowering the information content of feedback. However, both field data and laboratory data also show evidence of costly altruistic punishment. In the field data, a withdrawal request strategically supported with feedback retaliation (such that in theory, both parties face incentives to withdraw) is no more successful than a withdrawal request that is not backed by this threat. In the laboratory, cooperators emerge who, after receiving negative feedback that is purely strategically motivated, are likely to be much more emotionally distressed and are more likely to punish offending sellers' attempts to enforce feedback withdrawal, insisting instead on the sellers "making good." Strategic retaliatory claims hence appear ineffective, at least when applied to this group of cooperators, which in turn supports the efficiency of the conflict resolution system.

Similar costly punishment can be observed in many social settings, ranging from open disapproval of people who do not tip waiters or who do not wait patiently in a queue, to consumer boycotts of companies that use child labor or flout environmental norms.

Other studies have found evidence for crowding-out effects. Falk and Kosfeld (2006) found that tightening contractual obligations to cooperate may actually lower overall cooperation. Mellström and Johannesson (2008) studied whether providing financial incentives crowds out participation in blood donation and found evidence of partial support for this conjecture.

Our proposal is that market design efforts to change institutional settings in a group, with the goal of lowering exploitation, should be applied in such a way as to avoid crowding out the incentives of the group's preexisting counterforces to exploitative behavior.

What are the relevant questions to ask if one wished to pursue this proposal? We suggest that they include the following: Who has information about exploiters? How costly is direct punishment implemented by those with this information? And, more subtly, does the social norm being violated by these so-called exploiters truly benefit the group as a whole?

If the information about exploiters is only known locally and is not verifiable at a more aggregated level, and if the cost of punishment is low, then spontaneous punishment via a social norm is probably more efficient than a formal institution. When the social norm itself is not efficient, which can occur if a group's belief about what constitutes the social good is simply wrong, then there would seem to be a role for a group information institution that is able to assess claims about the benefit of this or that behavior in which members may or may not be engaged. For example, such an institution could verify not only whether a company truly has used child labor, but also whether that is indeed a bad thing to do given the actual circumstances faced by the employed children (e.g., if the alternative is child prostitution, then child labor may be the better option). Standard economic arguments on efficiency and transaction costs would apply to this design problem, inasmuch as the role of institutions would be to provide group mechanisms for information dissemination and punishment only when there is a returns-to-scale argument supporting such a role. By contrast, when there are decreasing returns to scale, such as when information is local and nonverifiable, the role of institutions would be more to give official group approval to low-level punishment rather than hindering or channeling it.

Imperfect Monitoring

Most of the current literature on the effectiveness of institutional settings in combating exploitative behavior assumes that each agent can perfectly observe the actions of every other agent. Yet the existence of perfect information in this area, like the absence of frictions elsewhere in an economy, is unrealistic (for further discussion, see Frijters and Foster 2013:70). In the real world, actions are not perfectly observed, and observing the consequences of actions does not always lead to straightforward conclusions about the underlying actions themselves or the intentions behind them.

To formalize this argument, we can think of x_{-i}, the actions of other players in

$$Y_i = f_i(x_i, x_{-i}, \alpha_i, \alpha_{-i}),   \qquad (10.6)$$

as being only imperfectly observed by agent i. All that is outwardly observed is a signal s_j drawn from a distribution around each true x_j. Punishment through institutions as well as the determination of future behavior in response to current experiences will then be based on s_j rather than on x_j. This may lead to errors—specifically, the punishment of investors (type I) or the nonpunishment of noninvestors (type II). The risk of either error may result in less prosocial behavior.

Ambrus and Greiner (2012) and Grechenig et al. (2010) tested the effect of noise in public good environments experimentally. In Ambrus and Greiner's 2012 experiment, decisions about whether to invest in the public good are binary (as in classic prisoner dilemma games), and noise is introduced as pure type I errors, through setting a 10% chance that any given investment is publicly shown as a noninvestment. This small amount of noise has a stark effect: investments and payoffs significantly decline and observed punishment increases. The reason seems to be that while group members punish the same way as they do in perfect monitoring conditions, punished investors react adversely and reduce their investments in the next round. This sets off a cycle which ends, after a while, in no investments in the public good by any group member. Essentially, the monitoring problem directly results in the crowding out of prosocial behavior. Grechenig et al. (2010) obtain very similar results, in an environment in which investment decisions are continuous and noise is added as a random shock on the investment of a group member.

Thus, imperfections in monitoring other group members pose a challenge to any attempt to sustain cooperation. Again, the logic sketched above would seem to apply: How might we design institutions such that punishment occurs at the level of those who have the best information?

An additional element comes to the fore in the presence of imperfect monitoring, which is the role of open examples. If monitoring is very difficult, the possibility arises to invest a lot of resources to get clarity on a small set of individual possible cases of exploitation, and then to have excessive punishment for small transgressions if the uncertainty is resolved. This design strategy again reflects basic economics: when it is hard to observe wrongdoing and hence when the odds of verifiable detection are low, the punishment of that wrongdoing when detected should be extreme to provide incentives to cooperate. A good example of this is the social demand made of politicians and judges to be beyond reproach: even small acts of criminality (say, stealing five dollars) would cost them their jobs, which are worth millions (for further discussion of the mechanisms that support this type of social dynamic, see Foster and Frijters 2016).

Conclusions and Frontiers for Market and Institutional Design

The three complications of the dilemma of exploitation that we have discussed are not the only ones, yet they are important and nontrivial to address. A market designer attempting to improve cooperation in a real-world dilemma situation will almost surely encounter these issues, and hence we propose that their study is worthy of market design economists.

The fact that exploitation is often carried out through the activities of subgroups poses the problem that market and organizational interventions targeted at a subgroup may also affect the whole group, and vice versa, both in terms of supporting and preventing cooperation. Reducing communication possibilities for firms in a market, for example, may be effective in preventing collusion, but may also reduce overall market transparency and thus lead to a second-best outcome in terms of market efficiency. Similarly, establishing a well-functioning mobile phone infrastructure in a developing country does not only support economic development, it may also improve coordination among militia groups. These examples as well as the results of Foster and Greiner's study suggest that communication is a key factor in within-group cooperation and the creation of an elite coalition. The ability to manipulate communication channels and information flows selectively for different subgroups is arguably a very important tool for an institutional designer.

In addition, an advanced understanding of the inner workings of group decision making will be important. Decision-making settings, such as whether everyone has a voice or whether majority vote determines the outcome, will have implications for institutional designers: What is the best way to sabotage cooperation within rogue subgroups to cause them to break up? In a setting that leaves it to group members to invent freely a means of making decisions, Ambrus et al. (2015) found that in a social decision-making context, extreme opinions are suppressed, and only intermediate and moderate individual opinions have a significant impact on the group's decision. Moreover, in decision-making settings that feature uncertain returns, the most risk-averse individual also has influence. This suggests that majority voting (which implies that only the median opinion in a group matters for a decision outcome) is not necessarily the best model of how groups make decisions; however, changing only the opinions of extreme members may have limited effects on the group's (pro- or antisocial) decisions.

The literature on cartels provides other hints about optimal structures to prevent or stem exploitative behavior by subgroups. Whistle-blower regulations allow members of a cartel to go free when they provide evidence about illegal behavior. In a similar way, other subgroups that exploit the larger group might be broken up by providing incentives to individual group members to stop cooperating within the exploitative subgroup. Our current thinking is that preventing coalition formation will probably be much easier than breaking up established coalitions and for this reason, new institutions and markets should be populated as much as possible with individuals who are not already in coalition with others and have limited incentives to form such coalitions.

Existing, endogenously evolved resistance against exploitation (e.g., in the form of altruistic punishers) may need active support through market design settings to be most effective. For example, reducing the costs of punishment or increasing its effectiveness may lead to a more powerful threat against antisocial actions, and thus to more cooperation and less eventually executed punishment (for related results, see Ambrus and Greiner 2012). On the other hand, the introduction of a police force, for example, might crowd out the motivation of effective vigilantes, and thus lead to a less-preferred social outcome if the vigilante organization was more effective than the police (e.g., due to local information advantages). An institutional designer needs to take such potential effects into account. As a general rule, the key issue is whether there are returns to scale in the aggregation of monitoring and punishment, or whether information and punishment are most efficiently collected and implemented, respectively, at the local level.

Finally, imperfect monitoring poses problems of its own. When punishment opportunities are provided to a social group but group members' actions cannot be observed perfectly, then the possibility of type I and type II errors in punishment arises. As Ambrus and Greiner (2012) show, unfairly punishing contributors may initiate group dynamics that lead to a decline in overall cooperation. Institutional design must address these issues. One way of doing this might be to allow the aggregation of noisy signals across group members, if this improves the quality of the information used in decisions about punishment. Another option is to increase punishment for small transgressions, or to put a lot of effort into reducing the uncertainty around a small set of suggested cases.

We close our discussion by asking what can be learned from this line of inquiry in regard to cooperation in other species and implications for "ecological design." As a preamble, we should mention that the economic concepts of free riding and contributing are not quite the same as the concepts of scrounging and production in animal studies, where the typical producer discovers new feeding grounds and the scrounger tags along and eats part of the discovered resource. Only when scrounging is seen as a persistent tactic and the scrounger and producer are not a genetic team (i.e., close kin) would it fit our formulation that in a utility sense the scrounger is free riding and the producers are contributing. When individual animals take turns at scrounging, the situation may in fact be an optimal form of sharing information and highly cooperative. Our comments below, therefore, refer to persistent behavior of an individual or (sub)group, rather than efficient specialization.

Much of the literature in evolutionary biology and animal social behavior takes it as given that both cooperation and cheating will be observed in many social group settings. The economist's singularly interventionist reaction to this reality can potentially bring new insights to the "ecological design" endeavors of those interested in changing the balance of cooperative and exploitative behavior in animal groups. In some cases, the exploitative behavior observed in animals may be seen as bad for humans; for example, when it threatens the survival of whole groups (species) that are endangered or otherwise already stressed. In other cases, humans may wish to augment the naturally occurring exploitation in competing species (e.g., as a means of fighting the spread of undesirable microbial colonies) by pushing them toward self-destruction through the selection of ecological design settings which encourage antisocial behavior within these groups.

Altruistic behavior in many animal species has been found in a number of studies to be genetically coded (e.g., Giraud et al. 2002; Dimitriu et al. 2014), although genetic coding of options may not be required in animals provided with brains that make decisions. Indeed, as pointed out by Dubois et al. (this volume), behavioral ecologists adopt the "behavioral gambit" which assumes that decisions made by nervous systems replicate the evolution of genetic alternatives. With one exception, discussed below, a genetic basis for human altruism is not (yet) a finding, and is not often even hypothesized, in the social sciences; rather, interpersonal differences in levels of cooperation within human groups are typically seen as the outgrowth of differences across people in their cultural programming or in the personal incentives they face. In line with this perspective, our market design approach to supporting cooperation disregards the extent to which human cooperation may be (at least to some extent) genetically coded, since such coding cannot realistically be manipulated or even observed by benevolent designers of institutions intended to promote cooperation. Instead, we focus on those differences in cooperation, either across or within groups, that arise due to differing individual or group incentives; that is, where individuals show *plasticity* in their behavior conditional on circumstances of the environment. It is those incentives and circumstances that are most effectively altered via strategic market design.

The one exception mentioned above is the hypothesis that genetic kinship may increase cooperation among individual groups or subgroups due to basic evolutionary concerns. This "kin selection" hypothesis, originally suggested in the biological sciences (see Wilson 2005), has been tested in human groups and has found some support in research into animal cooperation, such as the cooperation of males in polyandrous tamarin groups who share the care of babies that might not be theirs (Díaz-Muñoz 2011). However, kin selection is debated even in the biological sciences (e.g., see Mathot and Giraldeau 2010). Our concern here is not with genetic sources of cooperation—expressed via kin selection or in other ways—but with cooperation that is manipulable due to the opportunistic capabilities of the species in which it occurs.

In research programs aimed at slowing the growth of bacterial colonies, for example, ecological designers have thought about how to slow down the spread of cooperative genes (as in Dimitriu et al. 2014) or to provide survival advantages to antisocial genes. Our work suggests potential value in also considering how to crowd out cooperative behavior by individual organisms, change the level of noise in signals about socially relevant behavior, and/or encourage the formation of subgroups that cooperate only within themselves. One might achieve the crowding out of prosocial behavior in species that respond prosocially to cues about ambient levels of cooperation by manipulating the signals about levels of cooperation from surrounding individuals (a possibility implicitly suggested by Allen et al. 2016). In other species, crowd out might instead be achieved through the design of local environments in which public goods are already provided in a metered fashion, and hence where the incentives for individuals to produce them cooperatively to ensure survival are reduced. The logic of this approach is illustrated in the health example (Thomas et al. 2012) wherein cancer cells are triggered to become more aggressive (i.e., to exhibit more prosocial behavior, threatening the host) when they are deprived of oxygen—an input necessary for survival, if not exactly a public good. In humans, the level of crowding out of prosocial behavior and the formation of subgroup coalitions are manipulable in the short run due to the adaptability of the human species to environments with different information or norms. To the extent that nonhuman animals are able to adapt to changing environments (termed "plasticity of behavior" in the biological sciences), there may be scope for the discovery of particular mechanisms, such as those suggested above, to achieve short-term changes in crowd out or subgroup formation in such species. In the long run, genetic selection—whether engineered in the lab or invoked in the field through ecological manipulation—is an alternative strategy.

In mammals, coalition formation in response to circumstance has been observed (for a review, see Johnstone and Dugatkin 2000), and scientists have developed models of why these coalitions arise under certain ecological conditions (for a review, see Mesterton-Gibbons et al. 2011). This suggests that in some animal colonies, it may be possible to alter the costs of altruistic punishment and/or the likelihood of subgroup formation in the short run via ecological design. For example, Dugatkin's model suggests that the incidence of coalition formation in primates can be manipulated via changing the amount of resources under competition, the degree of credible exclusion of losers from those resources, and the level of individual investment required to create the coalition (Dugatkin 1998). Such manipulation might in practice consist of altering the distribution of introduced resources among group members; for example, by withholding resources from would-be coalition formers or providing them to individuals in possession of good-quality local information. Other levers that ecological designers might consider, guided by the economic approach, include breaking up existing coalitions through the temporary removal of key members of elite subgroups, or manipulating signal quality—perhaps via the installation of dummies or distractions into the environment.

Acknowledgments

This chapter greatly benefitted from comments by Sam Brown, Luc-Alain Giraldeau, Kiryl Khalmetski, Julia Lupp, Claus Wedekind, Bruce Winterhalder, and discussions with participants at the 2015 Ernst Strüngmann Forum on "Evolutionary and Economic Strategies for Benefitting from Other Agents' Investments."

11

Behavioral Consequence of Exploitation

Frédérique Dubois, Philipp Heeb,
Sasha R. X. Dall, and Luc-Alain Giraldeau

Abstract

This chapter addresses the behavioral consequences of individuals (exploiters) that use the investments of others (investors) rather than investing time or effort in procuring a resource themselves. The optimal exploitation strategy has been traditionally studied in behavioral ecology using the producer–scrounger (PS) model, a simple evolutionary game theoretic model in which producers (investors) search for resources while scroungers (exploiters) use the resources found by producers. For simplicity, a key assumption of the PS model is that the producer remains passive toward scroungers. As the presence of scroungers is costly, both empirical and theoretical evidence is reviewed that one major consequence of having exploiters is the adoption by producers of strategies that reduce the benefits of scroungers, giving rise to countermeasures by scroungers. In addition, scroungers have effects on population structure, notably by generating consistent differences among individuals and affecting spatial preferences within groups. Finally, although the PS game reviewed here is set in an explicit social foraging context, it is argued that it can be generalized to a large number of situations of social exploitation. Reviewing the impact of scrounging on populations should help generate parallels to explore the consequences of such exploitative behavior in economics and public health.

Introduction

Just about any fitness-enhancing behavior that requires time and effort is open to exploitation—usurpation of the benefit by an individual who has not invested the time and effort to attain that benefit. Having offspring, for instance, requires considerable effort, especially in species that invest in parental care. Building a nest as well as feeding and protecting the young are costly activities. In egg-laying species, such as birds and insects, some individuals avoid these costs altogether by depositing their eggs directly in the nests of a conspecific. Female starlings, for instance, are known to engage in this form of brood parasitism (Andersson 1985). Grasshoppers and crickets are known for their choruses in which males stridulate species-specific songs to attract mating females. The same is true in many species of frogs. Calling, in both cases, is energetically costly and exposes the caller to predators. Not surprisingly, there are reports for both insects and frogs in which noncalling silent individuals position themselves close to a calling male and intercept females on the way to the caller. These so-called "satellite males," which "orbit" around the territorial boundary of a calling male, obtain mating opportunities by exploiting the calling efforts of another male. When a couple of lionesses successfully hunt down a wildebeest on the Serengeti Plains, they are soon joined by a few other females that partake of the quarry without having participated in the effort required to capture the prey. They, too, are taking advantage of the hunting effort of their conspecifics.

    All of these cases are examples of exploitation, and there are many more in both animals and humans. It may seem surprising that an animal as dangerous as a lioness would do nothing to repel an exploiter that has come to eat her catch. Shouldn't she simply fight and attack the intruder to protect the precious resource that she just acquired? Similarly, should a female victim of a brood parasite do something to protect herself against this damaging exploitation of her parental effort? In this chapter we address the behavioral consequences of exploitation in terms of the individual that is exploited (i.e., the investor), the host, and the exploiter. We frame our discussion within evolutionary game theory using the producer–scrounger (PS) model. The game is set in an explicit social foraging context but we hope that it might be generalizable to all other systems of exploitation.

Origin of the Producer–Scrounger Game

When animals forage in a group in search of food that is distributed in discrete patches, one can assume that all individuals search for a patch and that, once one is discovered, all other group members will approach and obtain a share. This way of viewing group foraging is known as "information sharing," because the information about a location of a discovered patch is shared with all other group members. Information sharing constitutes a form of mutual or reciprocal exploitation: no member can gain anything by refraining from joining another individual's discovery, since every other individual will join when it discovers. Thus, under this scenario, all individuals engage in exploitation, and if the group size is G, the frequency of joining the population is $(G - 1)/G$.

    While observing the social foraging behavior of house sparrows, Barnard and Sibly (1981) quickly discovered that not all individuals in a flock invest equally in searching (investing) and joining (exploiting) behavior. So, instead of information sharing, Barnard and Sibly (1981) proposed that the searching and joining problem within a flock be analyzed as a two-option *n*-player evolutionary game in which the strategy "producer" corresponds to searching for food, and the strategy "scrounger" joins once a producer has been detected to find food. The game assumes that the payoffs to the scrounger are negatively frequency dependent: when scroungers are rare, they receive more than producers, but when scroungers are common, they receive less than the producers. This assumption means that the payoff curves of producer and scrounger will intersect at some frequency, when payoffs to both producer and scrounger are the same (see Figure 3.1 in Burton-Chellew, this volume).

The Evolutionarily Stable Strategy

One of the main differences between evolutionary game theory (i.e., the application of game theory to biology) and the traditional economic approach to game theory centers on the notion of evolutionary stability. Traditional game theory is based on the interactions of rational decision makers (i.e., humans). In evolutionary game theory, the payoff is a surrogate for Darwinian fitness, and strategies evolve over many repeated iterations of the game to reach the evolutionarily stable strategy (ESS), defined as a strategy that cannot be displaced by the occurrence of an initially rare mutant strategy. This can be illustrated intuitively with the classic hawk–dove evolutionary game: the hawk is an escalating strategy that quits only when injured, whereas the dove is a nonescalating ritualized fighter that concedes victory without injury as soon as the opponent initiates a fight. Although everyone in a population would fare much better if the population were composed solely of doves, this is not an acceptable solution in an evolutionary game because a mutant hawk would certainly win every fight against each dove it encountered. The hawk would clearly have higher fitness and thus spread throughout the population, eventually replacing the dove strategy entirely. Being a dove alone, therefore, is not an ESS and hence not an expected solution to the game.

In the case of the PS game, the scrounger strategy, by itself, is not an ESS. In a population comprised solely of scroungers, it is easy to imagine that if no one searches for food, then all will die of starvation. Likewise, a population made up solely of producers does not provide an ESS solution either because the game assumes that when a rare scrounger mutant arises in the population, its payoffs will be greater than those to producers; hence scroungers increase in frequency within the population. Because the game assumes that the payoff curves to producer and scrounger intersect, the number of scroungers increases until it reaches that equilibrium frequency where payoffs to both are equal. At that point, natural selection is stuck at a stable point, and the equilibrium combination of producers and scroungers is now a mixed ESS.

The Behavioral Stable Strategy

The argument for the PS game and its ESS solution are based on genetic change operating between generations. However, many of the behaviors of interest in this volume are not genetic, and the alternatives to "host" and "parasite" (investor and exploiter, respectively) are not necessarily genetically coded. In addition, most systems of interest reach solutions within the lifetime of the individuals. Thus selection is not required to act between generations. In the past, behavioral ecologists have not focused on these types of issues and have applied evolutionary game theory and its ESS solutions to all systems indiscriminately. This is due to an assumption that natural selection has equipped the brains of animals with learning rules which are the quickest at allowing individuals to reach outcomes that are equivalent to evolutionary solutions or the ESS (Harley 1981; Maynard-Smith 1982). We have referred to this credo as the *behavioral gambit* (Giraldeau and Dubois 2008), an assumption that learned outcomes of games will be those that match the outcome of evolution by natural selection. Until disproven, we suggest that it would be wise and more precise to distinguish evolved solutions from those achieved by behavioral decision by calling the former an ESS and the latter a behaviorally stable solution (BSS).

Up to now, the PS game (and its solutions) assumed that the producer remains passive in the face of the scrounger. Moreover, it failed to indicate whether scroungers should attempt to usurp the whole resource or simply part of it. In the real world, however, producers and scroungers are rarely passive, friendly companions. In some birds, like juncos, the scrounger aggressively displaces the producer from the discovered food source and takes it all. In many species of gulls, the scrounger aggressively pursues the resource holder until it drops the prey, generating impressive aerobatic flights. Lionesses tolerate each other at a kill but a lion does not. Hence there is a great deal of behavioral richness in real-life "social parasitism" (i.e., exploitation) that is not captured by a simple PS game. Below we explore how the scrounger strategy has given rise to a series of countermeasures. Our hope is that parallels can be generated to explore the consequences of such exploitative behavior in economics and public health.

Adaptations to Reduce Theft

Hoarding

The risk that a resource discovered by an investor is usurped by an exploiter is particularly high when there is a delay between the moment of discovery and use of the resource. This situation occurs in several bird and mammal species that store food for later use, hence leaving it unattended for several weeks or even months. In food-storing species, animals have developed different strategies to diminish the probability that their food is discovered and to mitigate the negative consequences that a pilferer may have on their fitness. More precisely, many food-storing animals scatter hoarded food items throughout their habitat, instead of accumulating them in a central place (e.g., a cavity in the ground or a burrow), as chipmunks and many other rodents do (Elliot 1978). Although scatter hoarding requires spatial memory abilities to allow individuals to retrieve food from a large number of locations (Shettleworth 1990; Pravosudov and Roth 2013), this strategy reduces both the risk of major loss (Vander Wall 1990; Jenkins et al. 1995) and the probability of pilferage, because dispersed food caches provide lower concentrations of food and hence are less attractive to other animals compared to central places. To avoid others learning the location of their food caches, individuals frequently modify their caching behavior when in the presence of observers, notably by choosing caching sites that are away from the observers and out of sight (Henrich and Pepper 1998; Lahti et al. 1998; Clayton et al. 2001; Bugnyar and Kotrschal 2002; Dally et al. 2005; Leaver et al. 2007; Shaw and Clayton 2013).

Aggression

Conversely, when food is consumed immediately or within a very short period of time after discovery, the owner of the food patch (the investor) may behave aggressively toward "joiners" (exploiters) by chasing them away from the patch, thus securing exclusive access to the food. Empirical evidence indicates that individuals seem to switch from peaceful sharing of food to aggressive encounters depending on ecological factors, such as the value of exploited patches or their level of spatial or temporal predictability (Goldberg et al. 2001). From a theoretical point of view, the question of which competitive tactic should be used has been frequently addressed by game theoretical models, most notably the hawk–dove game (Maynard-Smith and Price 1973); for an alternative approach based on economic conflict theory, see King et al. (this volume). The hawk–dove game considers a pair of opponents that compete for a resource using either an aggressive (hawk) or a nonaggressive (dove) behavior. It predicts that hawk is an ESS when the value of winning over a food patch (v) exceeds the cost of losing an aggressive encounter (c). Otherwise, the solution is a mixed ESS. The dove strategy, therefore, can never exist as a pure ESS, although nonaggressive resource sharing has been reported for a large number of animal species.

Since its original formulation, several variants of the hawk–dove game have been developed to generate more realistic predictions that apply to the process of foraging in groups, as will be reviewed here. One important deviation from the usual hawk–dove game is that in a foraging group, the two contestants do not arrive simultaneously at a food patch. The first to arrive (i.e., the finder) can thus exploit a patch before a joiner arrives, thus gaining a finder's share. This role asymmetry makes aggressive appropriation less profitable as the finder's share increases: the remaining food becomes insufficient to cover the cost of fighting, and the joiner retreats in the face of conflict escalation (Dubois et al. 2003). The finder should always compete aggressively when opponents are equal in their fighting ability, regardless of the size of the finder's share (Dubois and Giraldeau 2003, 2005; Dubois et al. 2003). Even if they could do better by sharing the remaining food, the first to decide on a strategy (i.e., the finder) should never play dove, because the joiner would then systematically play hawk and get all the remaining resource at no cost. Sharing could arise, however, if a finder that is prone to sharing, and hence to playing dove, changes its decision when its opponent decides to play hawk. In this case, food sharing should occur notably when food patches are of intermediate value, or when there is uncertainty about the value of the remaining resource (Dubois and Giraldeau 2007). Food sharing should also be more common when resource distribution is heterogeneous (e.g., nonrandom distribution of food patches, food patches of variable quality), as heterogeneity likely increases joiner uncertainty about the availability and quality of food patches. Accordingly, higher levels of aggression in nutmeg mannikins (*Lonchura punctulata*) have been reported when birds had no information about the value and location of the food patches (Dubois and Giraldeau 2004).

Another reason why the pure dove strategy is never an ESS is because hawk–dove games typically consider only a single contest. Most group foraging situations, however, involve opponents that interact repeatedly over discovered food patches. To counter this, Dubois and Giraldeau (2003) developed an iterated hawk–dove game which predicts that animals should share the resources without aggression when they face situations analogous to a prisoner's dilemma. Specifically, when the cost of escalation is relatively small, mutual dove provides a higher payoff than mutual hawk. Here, however, a dilemma emerges: playing hawk against a dove results in maximum payoff while playing dove against a hawk results in maximum losses. Under such conditions, playing hawk is the best strategy for both contestants if they interact only once. If they interact repeatedly, however, a conditional strategy such as tit-for-tat (i.e., playing dove on the first round and then doing what the opponent did on the previous round) can guard against the invasion of hawk players. The probability that food patches are peacefully shared between the two contestants depends on, among other parameters, the finder's share (i.e., the fraction of the patch that can be exploited exclusively by the finder before a joiner arrives). More precisely, the rate of aggression is expected to increase as the finder's share decreases, and then should increase with patch quality. Given that increasing competitor numbers likely reduce mean distances among individuals, the amount of food that can be gained by the finder before a joiner arrives should decline. Hence the benefits of defending increase with the number of competitors. As such, aggression should be more frequent in large group sizes (Dubois and Giraldeau 2003).

Another unrealistic assumption of the original hawk–dove game is that food patches are always contested by two opponents. In many instances, however, a large number of individuals converge at the patch, thereby forcing the finder to defend its resource against a large number of intruders. In this case, greater competitor numbers increase not only the benefits but also the costs of defense, resulting in a dome-shaped relationship between the frequency of aggression and the density of competitors (Dubois and Giraldeau 2003). Experimental evidence indicates that the aggression rate peaks at intermediate group sizes (Jones 1983; Goldberg et al. 2001). Dubois and Giraldeau (2003) assume that all intruders should always attempt appropriation of an owner's resources and hence analyze only whether appropriation should be aggressive or not. The question of whether a resource is worth defending, however, is relevant only when the resource obtained by a competitor is worth appropriating. To address this issue, Dubois and Giraldeau (2005) developed a game theoretical model that incorporates both the producer–scrounger and hawk–dove games, to explore how the interaction between appropriation and defense generates patterns of aggression in resource patches. Like other variants of the hawk–dove game (Sirot 2000; Dubois and Giraldeau 2003; Dubois et al. 2003), their model predicts that the frequency of aggressive interactions should decrease as the encounter rate with food patches increases. However, the predicted decrease in aggression with increasing patch density does not result from a decrease in individuals' level of aggressiveness but from a decrease in the proportion of scroungers whose aggression levels remains somewhat constant. Similarly, Dubois and Giraldeau's model (2005) predicts a dome-shaped relationship between group size and aggression frequency, as does Dubois et al. (2003), but the number of competitors has almost no effect on aggressiveness.

Finally, because of the finder's share, finders can gain more from a resource than joiners. Therefore, hawk–dove games which apply to foraging groups typically predict that the finder should always compete aggressively unless its fighting ability is well below that of its opponent(s). When engaged in defense, however, the finder would be incapable of preventing "distraction sneakers" (i.e., individuals that employ distraction so as to sneak into an area) to gain access to a resource left temporarily unguarded, leading to a reduction in the benefits of defending. The presence of sneakers should affect the expected level of aggression within groups. Supporting this idea, Dubois et al. (2004) demonstrated that introducing a distraction-sneaking tactic into the hawk–dove game decreases the expected proportion of aggressive animals playing hawk, particularly when sneakers search for both unchallenged resources and opportunities to appropriate food patches.

Ganging Up

When the resources are not defendable by a single individual, the finder may attract other individuals in an effort to form coalitions, thus increasing its chance of keeping the resources and minimizing the risks of aggression and injury. This type of strategy is mostly used by young or low-ranking individuals who benefit from "ganging up" to evict competitors with higher fighting ability (Heinrich and Marzluff 1991). Coalitions may also prevent interspecific kleptoparasitism. In particular, in social carnivores, one major argument for the sociality of lions has been to be able to defend their prey against theft by hyenas. Similarly, hyenas form large raiding groups to dislodge lions from their captures (Kruuk 1972).

Thus, whenever it is in the interest of finders to recruit others at resource patches, recruitment signals can be expected to evolve. For instance, when juvenile ravens find a carcass in winter, they call to attract others to form a group. This group of recruits is able to keep territorial owners at bay and consume the food (Heinrich and Marzluff 1991). Thus the decision to produce recruitment calls depends on the possibility of obtaining a sizable share of the resources. In a seminal paper, Elgar (1986) showed that house sparrows mostly produced recruitment calls when resources are divisible among the users. The use of recruitment calls to attract recruits at the resource patch has been proposed to play a role in the establishment of certain animal associations (Richner and Heeb 1996).

Consequences for Spatial Structure of Groups

When foraging in social groups, the spatial position of individuals within these groups is not neutral. The costs and benefits of individuals in groups may depend on their spatial position. Individuals using a producer or scrounger strategy might benefit from adopting specific locations. Barta et al. (1997) examined whether the strategies of producer and scrounger played by an individual should affect its spatial position within a group and the geometry of the foraging group. They found that the existence of the scrounger strategy leads to a decrease in the surface area occupied by foraging groups, and the average distance from the center of the groups or the average distance to the nearest neighbor was smaller. Scroungers tended to occupy positions in the middle of the group, close to other subjects, whereas producers tended to occupy the periphery. The authors concluded that groups containing producers and scroungers should be more compact compared to an equivalent group made up solely of producers. They argue that scrounging may be an alternative (or additional) factor in the promotion of the dominants' known preference for central positions (Barta et al. 1997).

However, the distinct positional advantages of each strategy suggest that an individual alternating between optimal producer and scrounger alternatives would likely incur a cost associated with a shift in spatial position within the group. Depending on their value, these costs would lead to the stabilization of specific strategies in given locations. The spatial location of an individual within a foraging group likely affects the payoffs anticipated from using the producer or scrounger tactics. A producer, for example, profits most by being furthest from any others, reducing competition by fellow producers while increasing the time required by scroungers to reach its discovery. A scrounger, on the other hand, benefits most by being as close as possible to all potential food finders, minimizing the time elapsed between the food discovery and its arrival at the patch.

Later studies examined the role of PS strategies in explicit spatial scenarios. Flynn and Giraldeau (2001) examined whether the exploitation of companions' food discoveries had spatial consequences for the foraging individual. Specifically, they examined the association between tactic use and the spatial characteristics of foraging flocks of nutmeg mannikins (*L. punctulata*) as a model organism. They predisposed some individuals toward the producer tactic by pretraining them to find food hidden under lids. They also predicted that in dominance-structured PS games, subordinates should prefer to play producer (because their position in the hierarchy prevents them from getting food by scrounging) and place themselves on the periphery of the groups. Being at the center of groups provides the possibility of accessing more information, and scroungers are expected to be more numerous. Centrally located individuals must scan to detect scrounging opportunities. The study of vigilance behaviors in foraging groups has received quite a lot of interest from biologists. The association between scrounging and vigilance suggests that scanning frequencies reported within central portions of foraging groups may have to do with the use of scrounger strategies, and probably less with predation hazard (Coolen et al. 2001). It follows that antipredatory vigilance behavior may be more directly measured in peripheral individuals whose scanning is unlikely to be related to the use of scrounger tactics.

In confirmation with earlier studies, Flynn and Giraldeau (2001) concluded that individuals who frequently use the producer tactic forage preferentially away from the center of the flock, whereas those that favor scrounger tactics prefer more centrally located positions. If this result holds in other scenarios, this implies that symmetric PS games will not apply in natural conditions, and that other models would need to be developed to take position asymmetries into account.

Exploiting Information from Others

In contrast to solitary animals, which generally must acquire information directly from the successes and failures of their own decisions, social animals are able to obtain information by sampling the environment themselves as well as by observing the decisions of their companions. These sources of social information add value to different alternatives (Danchin et al. 2004).

The degree to which animals will rely preferentially on social information appears to depend on the ease with which social and personal information can be simultaneously collected and the difficulty of gathering accurate personal information from the sampling of their environment (Templeton and Giraldeau 1995). When animals gather information in this way, the information provided by a sequence of social tutors can induce herd-like phenomena or "informational cascades": decisions are made regardless of the personal information at hand. In this scenario, individuals "blindly" copy the decision they witnessed (Bikhchandani et al. 1998). Although copying predecessors' actions often leads to the adoption of the correct decision, it is also prone to the adoption of incorrect decisions, imposing fitness-related costs (Bikhchandani et al. 1998; Giraldeau et al. 2002). Such costly cascades are expected to occur when the observation of companions is limited to their decisions rather than the cues on which these decisions were based (Bikhchandani et al. 1998; Giraldeau et al. 2002). Results obtained by Rieucau and Giraldeau (2009) provide experimental evidence that nutmeg mannikins (*L. punctulata*) tend to disregard personal information when social information is sufficiently convincing. These birds relied on social information more when the cue used failed to predict the location of the fast feeder. Their study provides the first experimental evidence in nonhuman animals that is consistent with the propagation of informational cascades observed in human crowds (Bikhchandani et al. 1998). Their results raise the issue that the use of personal information, independent of its quality, does not insulate individuals from the use of social information (Valone and Giraldeau 1993).

In nature, competitive situations can exist where it pays for the finder to hide or conceal public information. For example, the behavior or performance of one species can be used as a source of information about mutually exploited resources by putative competitors (Danchin et al. 2004; Goodale et al. 2010). Increased overlap in resource use may result in costs for the information source in terms of enhanced interference and exploitation competition, resulting in an "evolutionary arms race" that favors acquiring and hiding information (Seppänen et al. 2007). As an example, studies have shown that pied flycatchers (*Ficedula hypoleuca*) use great tits (*Parus major*) as a source of information in habitat and nest-site selection decisions (Loukola et al. 2014). The flycatchers can gain fitness benefits (by laying larger clutches) from the information they obtain concerning nesting sites, by observing the clutch size laid by the tits when nesting in proximity to great tits. In contrast, tits suffer from the association and resulting competition (they lay smaller clutches). In response to this form of "social parasitism," tits attempt to hide the information provided by their clutches by covering their eggs with different materials (e.g., hair, moss, moss sporangia, grass). After performing an experiment where the risk of having flycatchers as parasites were increased, Loukola et al. (2014) found that tits put more hair on the eggs and covered them more carefully. These results illustrate the fact that when exploitation is costly, it pays for the "producer" (investor) to reduce the amount of information available to the "scroungers" (exploiter).

When animals forage socially, individuals can obtain prey by using the behavior of others when they inadvertently provide information that food has been located. This inadvertent social information can be of two types (Danchin et al. 2004): it may provide social information simply by indicating the location of the resource, or it may provide public information to indicate the quality of the resource based on the performance of the individual already engaged in exploiting it (Valone 2007). For the individual using the information provided, public information is considered to be better than social information because it is used preferentially when it is equally costly to obtain as other types of inadvertent social information (Coolen et al. 2005). It would be interesting, however, to explore situations where public information provides no additional benefit to inadvertent social information (e.g., when resource quality has no variance or similar situations). Because scrounging, and hence the use of inadvertent social information, is mutually incompatible with producing (Coolen et al. 2001), any increase in the stable equilibrium frequency of scrounging results in a decreased number of producers that are concurrently searching for prey, and thus in lower predator search efficiency. Therefore, prey may be expected to evolve characteristics that can induce high rates of scrounging in their predators to reduce predator search efficiency (e.g., prey crypticity; Barrette and Giraldeau 2006). Another such trait may be prey clumpiness: larger prey clump sizes are predicted to increase the stable equilibrium frequency of scrounging (Caraco and Giraldeau 1991) and have been demonstrated to reduce predator efficiency at finding patches (Coolen 2002). Predators increased their use of the scrounger tactic in response to increased average prey clump size.

In a simulation study, Hamblin et al. (2010) further tested some of the evolutionary outcomes of these predictions. As predicted, they found that as prey grouped together, the frequency of scroungers among predators increased and stabilized after prey reached a certain clump size. Surprisingly and contrary to expectations, their simulations showed that prey evolved toward the highest clumping against predators without social information. Prey evolved toward smaller clump sizes when facing predators with social and public information predators. Their study was the first to demonstrate how information use by predators evolves in response to prey–predator dynamics. The prey survived better when the predators used either social or public information, which shows that scrounging, and hence the use of inadvertent social information in any form, actually reduced predator efficiency. Consistent with this suggestion, predators reduced their investment in scrounging much more rapidly when they had access to public information compared with when they had access only to social information. This suggests that there would be an advantage to prey to evolve traits that reduce the ability of predators to provide public information while they are being exploited. It would be interesting to explore which traits are selected to reduce the availability of inadvertent social information.

The fact that behavioral strategies of predators, in particular scrounging, can affect the behaviors of their prey can lead to population regulation through their effects on individuals' reproductive rate and mortality. Coolen et al. (2007) explored the effects of scrounging on prey–predator population dynamics and showed that the presence of scrounging predators allows an increased predator population size and contributes to the regulation of both predator and prey populations. This result may have general value. For instance, if the prey in one context is a valuable resource (say a forest or a stock of fish), then one could envisage implementing measures to increase scrounging among the individuals engaged in exploiting it as a means of increasing the sustainable use of the resource (see Valone et al., this volume).

Spying on others as they go about their lives can have significant consequences. On one hand, those being spied upon become vulnerable to having their efforts exploited. However, in some contexts, the presence of spies can provide opportunities to control social interactions more effectively. This could select for those being spied upon to act in ways to minimize exploitation by information parasites, or even to manipulate spies through the information being observed. Either way, the availability of inadvertently provided information often changes the behavioral strategies we expect to evolve (McNamara 2013).

Guarding against Information Parasitism

It seems obvious that it is rarely in an individual's interest to have his behavior monitored, since this would put the individual at risk of having the fruits of his labors exploited. Clearly the adjustments discussed above in scatter hoarding (i.e., changes in caching behavior in the presence of conspecific observers to minimize the risk of pilferage) might be thought of as strategies to mitigate the risk of information parasitism. Nevertheless, putative pilferers can also adopt sophisticated counterstrategies once they know that their behavior is being monitored. For instance, common ravens will regularly cache excess food, and it is well known that such hoarders adjust their behavior in the presence of potential pilferers to mitigate the risk that their hoards are pilfered (Bugnyar and Heinrich 2005, 2006). This involves attacking known witnesses to their caching attempts or quickly retrieving caches in the presence of such individuals, depending on the relative social status of the individuals involved (Bugnyar and Heinrich 2005). Such strategic defenses have counterstrategies that involve the strategic use of social misinformation by the wannabe pilferers. For instance, putative pilferers of dominant hoarders will delay approaching known cache locations by cursorily searching the general areas around the caches (and thereby acting deceptively ignorant) only when the hoarder is present (Bugnyar and Heinrich 2006). This provides opportunities to pilfer the caches before the dominant bird has a chance to drive them off.

Such strategic deception in the presence of potential information scroungers has also been documented in nonfood-hoarding contexts. It seems to be a particularly effective ploy when the putative information parasites are dominant to the potential victims. For instance, in captive food retrieval experiments in chimpanzees, dominant individuals typically monopolize any food for which they know the location. In response, when subordinates have exclusive knowledge of some food locations but can only access them in the presence of dominants, they will engage in strategic maneuvering by waiting or hiding to obtain pieces of food or even proactively distracting the dominants by deceptively socializing with them to keep them away from the food (Hare et al. 2000). Thus, in at least some cognitively complex species, individuals can assess what conspecifics can or cannot see or know, and deploy some sophisticated strategies to counter the threat of being exploited by information parasites.

Being Watched Enhances Social Information

During social (including sexual) encounters, an individual's behavior is often influenced by the interactions of other conspecifics and the predictable responses that they make (Dall et al. 2004; McNamara et al. 2009; Bergmüller and Taborsky 2010; Schuett et al. 2010; Wolf et al. 2011). Thus, when animals interact socially, they can pay attention to each other's behavior to make better decisions. Once individuals use such basic social information, it changes selection and the day-to-day reinforcement of behavioral patterns over time in the presence of audiences. In some contexts (e.g., when competing aggressively for resources or interacting cooperatively), this can favor (both evolutionarily and behaviorally) individual behavioral differentiation (Dall et al. 2004; McNamara et al. 2009; Wolf et al. 2011). However, it can also make negative frequency-dependent payoffs to be adaptively flexible (Wolf et al. 2008; Dubois et al. 2010).

To see why, consider the hawk–dove game. In the basic game of competition over resources of value v, given that getting into an escalated fight costs $c > v$, the ESS (or BSS) is for a proportion v/c of individuals to assume the hawk strategy (always escalate if challenged) while $1 - v/c$ play the dove role (always capitulate without fighting) at any given moment. There are, however, two ways in which the ESS mixture of tactics in a population can be maintained by frequency-dependent payoffs alone:

1. Each individual can perform actions randomly with fixed probabilities, and thus generate the predicted mix of strategies in large populations.
2. Fixed proportions of individuals can play each strategy consistently.

The conditions favoring the evolution and maintenance of one or another of these forms of evolutionarily (or behaviorally) stable mixtures of behavior have yet to be elucidated in general, although investigating the adaptive dynamics of biological games may generate insights (Bergstrom and Godfrey-Smith 1998). Indeed, for the basic hawk–dove game in finite populations, such analysis suggests that populations of ESS (BSS) mixtures of individuals specializing in each strategy will always evolve toward monomorphic populations of individuals playing hawk OR dove randomly with ESS (BSS) probabilities (Bergstrom and Godfrey-Smith 1998). This is because stochastic variation in the frequency of hawkishness (e.g., due to demographic noise) within populations will penalize individuals that commit themselves to playing either strategy consistently (Bergstrom and Godfrey-Smith 1998).

However, adding the possibility that individuals might "eavesdrop" on one another (e.g., base their tactics on the outcomes of their opponents' last fights; Johnstone 2001) to the hawk–dove game with adaptive (e.g., replicator) dynamics favors the alternative outcome: strategies that generate consistent individual differences in aggression will be at an advantage in monomorphic populations in which all individuals play hawk and dove randomly at ESS (BSS) probabilities. This is because, with eavesdropping in the population, more consistent aggressiveness (high or low) is favored since, by being more predictable, individuals can avoid getting into extended (costly) fights. Moreover, with increased interindividual variation in aggressiveness, increased levels of eavesdropping will be favored to minimize the chance of fighting with the more aggressive individuals (who are more likely to have won their last fight), and so on. This dynamic feedback will eventually result in polymorphic populations that are composed of extreme types at ESS (BSS) frequencies, in which individuals are always either hawks, doves, or eavesdroppers (Dall et al. 2004). Thus, adding the possibility of being spied upon while fighting results in a qualitatively distinct evolutionary or behavioral outcome, since consistency can be favored when being predictable gets competitors to respond, in the future, to improve focal individuals' payoffs. Perhaps the fact that fights over resources rarely occur in social isolation can explain why consistent individual differences in aggressiveness, manifest in dominance hierarchies, are common in a wide range of species.

In general, it is possible that when individual behavior is being monitored, variation itself can promote further variation by favoring social information use. This is because the existence of stable interindividual variation means that there is something to learn from monitoring others, which in turn can favor individual differentiation among those being monitored (McNamara 2013). Indeed such intuition also holds in a model of trust and cooperation (McNamara et al. 2009), where allowing individuals to monitor each other's cooperative tendencies, at a cost, can favor polymorphisms in trustworthiness. This variation, in turn, favors costly "social awareness" in some individuals (McNamara et al. 2009). Indeed, feedback of this sort might explain the individual differences in trust and trustworthiness so often documented by economists in experimental public goods games across a range of cultures (e.g., Fischbacher et al. 2001).

Conclusion

The optimal frequency of exploiters and investors within populations has been traditionally analyzed using the PS game. Although several predictions of the PS model have been supported by experimental evidence, it is based on highly simplified assumptions. In particular, though producers are generally considered as passive competitors that do nothing to prevent scroungers from exploiting their efforts, they frequently adopt behavioral strategies aimed at reducing the benefits of scrounging. For instance, when competing for food, producers can use aggressive behavior to chase away the scroungers and get exclusive access to the resource instead of sharing. Similarly, individuals that cache food or sample their environment to select a suitable habitat to settle may modify their behavior when in the presence of others to reduce the risk of pilferage or information parasitism, thereby reducing the proportion of exploiters. Alternatively, although PS models consider all group members as equally effective in searching and joining, individuals may modify the costs and benefits associated with both strategies simply by adjusting their spatial location within the group to their strategy. Since scroungers have effects on the structure and dynamics of populations, notably by generating consistent differences among individuals or by contributing to the regulation of both predator and prey populations, it is important to develop more realistic models that include these behavioral adaptations. Such models might be useful to evaluate measures aimed at sustaining resources of economic value, for instance by manipulating predation pressure (i.e., the proportion of producers) on prey populations.

12

Exploitative Strategies

Consequences for Individual Behavior, Social Structure, and Design of Institutions

Andrew J. King, Michael Kosfeld, Sasha R. X. Dall,
Ben Greiner, Tatsuya Kameda, Kiryl Khalmetski,
Wolfgang Leininger, Claus Wedekind, and Bruce Winterhalder

Abstract

"Exploitation" or free riding are names for strategies by which agents benefit from other agents' investments. This chapter reviews the consequences of these exploitative strategies for individual behavior, social structure, and design of institutions. From an evolutionary perspective, it begins by outlining how natural selection should act to construct behavioral connections that maximize the benefits and minimize the costs of sociality for individuals. Individuals are predicted to show specific leaving or joining decision rules that will construct groups composed of complementary strategies; alternatively, they should be plastic in response to their social environment, which can lead to conditional strategies and social niche construction. What happens on an individual level impacts, in turn, social structures. When individuals have fewer or more frequent interactions with a set of specific (known) individuals, "groupiness" may result to reduce uncertainty in interactions. In humans, common knowledge of within-group norms can further facilitate coordination on socially efficient equilibriums and establish cooperation. Once groups are maintained and cooperate to produce and share resources, they become open to exploitation by other groups, which is directly relevant to the design of institutions. Economic conflict theory offers a potential framework for understanding and predicting exploitative behavior between groups. Through a better understanding of exploitation at these different levels, it is hoped that the payoffs of specific interactions can be adjusted to reduce the negative impacts on a system.

Group photos (top left to bottom right) Michael Kosfeld, Andrew King, Claus Wedekind, Sasha Dall, Kiryl Khalmetski, Bruce Winterhalder, Ben Greiner, Tatsuya Kameda, Andrew King, Wolfgang Leininger, Michael Kosfeld, Ben Greiner, Kiryl Khalmetski, Bruce Winterhalder, Sasha Dall, Claus Wedekind, Tatsuya Kameda, Wolfgang Leininger, Michael Kosfeld and Tatsuya Kameda, Andrew King, Michael Kosfeld

Introduction

"Social parasitism" (e.g., Cote and Poulin 1995; Safi and Kerth 2007), cheating (e.g., Velicer et al. 2000; Sandoz et al. 2007), or free riding (e.g., Hardin 1968) are all names for strategies by which agents benefit from other agents' investments. Our group was tasked with understanding the consequences of these exploitative strategies for individual behavior, social structure, and design of institutions.

The first obstacle that we faced involved identifying and quantifying exploitation. This is not trivial. Where individuals consistently invest differently in, or receive unequal rewards from, social interaction, counterstrategies tend to evolve which offset or mediate costs (Welbergen and Davies 2009; Kilner and Langmore 2011; Daugherty and Malik 2012). Thus, when we look at the outcome of social interactions across time and contexts, it can be difficult to quantify exploitation because individual behaviors and/or the social system have managed or mitigated it. In other words, there are often scenarios in which it is better for an agent to interact with those that exploit their efforts, or take more than their share, than not to interact with them and get none of the benefits of social interaction. Gaining some benefit is better than getting no benefit at all (West et al. 2006b, 2007c).

To overcome this barrier, we took a Hamiltonian[1] approach (see Figure 12.1) and quantified four types of social interaction (Foster et al. 2001). First, it is useful to distinguish the individual who initiates an interaction from the individual or individuals that it influences (of course there are always consequences for the initiator as well). The former is typically called the *actor* and the latter the *recipient(s)*. Next, assuming that any consequences of actions can be either positive or negative, Hamilton derived a classification scheme that enabled social interactions to be classified into four categories:

1. Mutualism, if the consequences are positive for both actors and recipients
2. Selfishness, if the consequences are positive for the actor but negative for the recipient
3. Altruism, if the consequences are negative for the actor but positive for the recipient
4. Spite, if the consequences are negative for both parties (i.e., actors pay a cost to inflict a cost on recipients)

The strict (original) classification scheme for these four types of social interaction envisages the consequences to lifetime fitness. Many evolutionary ecologists, however, prefer to think instantaneously about the consequences (i.e., accounted to some short-term proxy of fitness like net rate of energy gain)

---

[1] William Hamilton (1936–2000) was an English evolutionary biologist, widely recognized as one of the most significant evolutionary theorists of the twentieth century.

Effect on recipients

Effect on actors		+	−
Effect	+	Mutualism	Selfishness
on actors	− or 0	Altruism	Spite

Figure 12.1 Classification of social actions, the consequences of which can be either positive or negative.

for such interactions (Barta, this volume). Both views are relevant and applicable here. When juxtaposing the actions of different individuals toward each other in an interaction, mutualistic cooperation (mutual mutualism) can be distinguished from altruistic cooperation (mutual altruism). In a situation where mutual altruism is feasible (West et al. 2007b), we define the term "exploitation" as benefitting from the altruism of another individual without taking an altruistic action. From the perspective of altruistic cooperation, such an action is selfish, since it increases an individual's own payoff (or fitness) at the cost of another's payoff (or fitness). Note that this definition encompasses the producer–scrounger framework (Barnard and Sibly 1981; Vickery et al. 1991), reviewed by Barta (this volume), as well as social dilemma games (e.g., prisoner's dilemma or public goods games) that are studied in economics, reviewed by Burton-Chellew et al. (this volume). When there is no concurrent altruistic action taken by an individual (i.e., when *mutual* cooperation is not feasible), then a selfish action by an individual is not called "exploitative." Exploitation is defined as benefitting from the altruistic cooperation of others.

Now that we have defined selfish social actions that correspond to exploitation as broadly conceived, we can classify, compare, and contrast the consequences of these exploitative strategies for different levels of social organization. Each of these levels will be discussed in turn; for an overview, see Figure 12.2.

Consequences of Exploitation for Individual Behavior

Dubois et al. (this volume) present a broad overview of the behavioral consequences of exploitation and provide an excellent framework for investigating how hoarding (Andersson and Krebs 1978), aggression (Manson and Wrangham 1991; Clutton-Brock and Parker 1995), and coalition formation (Silk 1982) can be caused by the consequences of exploitation. Here, we focus on how the relative frequency of different types of social interactions that individuals experience (Figure 12.2) can play a critical role in their survival and reproduction. Specifically, how frequently and in which contexts an individual interacts with others, with either positive or negative outcomes, can influence the emergence and spread of cooperative behaviors (Gulati 1998; Ohtsuki et al. 2006), social learning (Franz and Nunn 2009; Hoppitt and Laland 2011), and directly transmitted diseases (Newman 2002; Keeling and Eames 2005; Mossong et al. 2008). Consequently, selection should act to construct

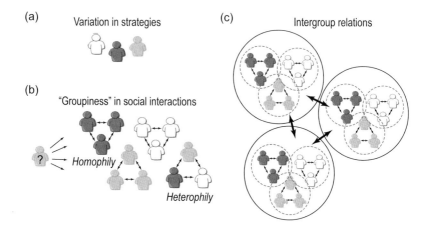

Figure 12.2 Consequences of exploitation for individual behavior, social structure, and the design of institutions. (a) Individuals exhibit between-individual differences in strategy. These strategies need not be fixed. For example, the "white" strategy could be to switch adaptively in accordance with the frequency of behaviors exhibited by the others. (b) Given that interacting individuals vary in their strategies, evolution should select for leaving or joining decision rules that either change group mean (and variance) in strategy or select for individual plasticity in strategy. Both mechanisms function to change strategies to match (homophily) or complement (heterophily) the group. This acts to reduce uncertainty in strategy space, which in turn affects group-level outcomes (e.g., the total amount of food a group acquires). (c) As a result, between-group differences in production and investment emerge, which can lead to intergroup conflicts. Note: populations can also be monomorphic for a strategy but still show individual differences in expressed tactics.

behavioral connections that maximize the benefits and minimize the costs of sociality. In practice this will mean associating with certain individuals and avoiding others; functionally, this will act to reduce uncertainty in their social interactions (Aureli and Vanschaik 1991; Sueur et al. 2011; Farine et al. 2015). Consider a schooling fish, which might interact with thousands of other fish over the course of a day, but tends to form disproportionately close associations with others that are similar to itself in size or activity (Hoare et al. 2000a, b). This may confer increased protection from predators or improve social coordination during migratory movements (Croft et al. 2009). Similarly, while a female baboon will live with many of the same troop mates from birth to death, she will tend to associate with baboons of similar competitive ability (King et al. 2009). This acts to reduce the potential for conflict and increase opportunity for cooperative interaction (King et al. 2008, 2011; King and Sueur 2011), and affects group-level phenotypic composition (Farine et al. 2015).

Group phenotypic composition—or more precisely, the heterogeneity of strategies in groups—can have important implications for individual success and fitness (Laskowski and Pruitt 2014; Farine et al. 2015). When individual strategies are fixed—individuals play pure producer or scrounger strategies (Giraldeau and Beauchamp 1999)—selection should result in the evolution of behaviors that will affect heterogeneity of groups—leaving or joining decision rules, or forceful eviction of particular members from groups (Stephens et al. 2005; Kerth et al. 2006; Kerth 2010). In contrast, if strategies are conditional (e.g., Fischbacher et al. 2001) or plastic (Furtbauer et al. 2015) and can be modulated in response to the others—individuals can express varying levels of producing and scrounging in different contexts (Morand-Ferron and Giraldeau 2010)—we expect that selection acts to produce plasticity in response to the social environment. This could lead to social niche construction, whereby individuals alter their associations and interactions with others to increase their chances of surviving and reproducing (Flack et al. 2006; Laskowski and Pruitt 2014; Farine et al. 2015).

Consequences of Exploitation for Social Structure: "Groupiness" and the Reduction of Uncertainty

Dubois et al. (this volume) review the consequences of exploitation for the spatial structure of groups, with a particular focus on the producer–scrounger game. Here, we focus on the consequences of exploitation for social structure (which may or may not be independent of spatial structures) and continue with the theme of flexibility (or plasticity) in strategies using a recent study of producer–scrounger dynamics undertaken by one of our group (Kameda). In this study, people were tasked with finding "treasure" hidden in a 5×5 grid, and groups of individuals ($N = 4$) independently decided whether or not to "dig" in one of the grid areas for a treasure, paying some search cost. Each search decision is independent and without communication, so that foragers cannot coordinate their search effort. Importantly, each group member bears the cost of search individually, but a treasure found will be shared evenly by all members. This setup can potentially yield a producer–scrounger situation (see Barta, this volume), where the group production function (i.e., the probability of the treasure to be found) increases monotonically with more cooperators (searchers) but diminishes in margin, because searchers cannot coordinate with each other about where to search (at best, they can only pick up one spot randomly). With feedback about net payoffs and the number of searchers after each round, a phenotypic polymorphism of producers and scroungers emerges. In the first ~33 rounds (except the first round), the producer frequency distributes in a unimodal, almost symmetric distribution. From round ~67 to the end of the experiment (round 100), the distribution of producing across individuals becomes U shaped: roughly 30–40% committed scroungers never cooperate, 20% committed producers almost always cooperate, and the remaining 30% are in-between. Kameda's team then reassigned the subjects to new groups as follows: the most cooperative players in each group were assigned to an "elite" group (the most cooperative team), the next most cooperative players into a second group, the third most cooperative players into a third group, and the least cooperative players into the fourth "delinquent" group. Then they played the game again. Although the average cooperation level (proportion of producing) was higher in the "elite" group than in the "delinquent" group, all newly formed groups showed a typical U-shaped distribution. This suggests that the division of producer–scrounger roles in a group is not guided solely by stable social preferences, but is (to some degree) plastic in response to the social environment and can emerge through repeated social interactions. The emergence of U-shaped distributions over time has also been observed in other types of collective tasks by human subjects (Kameda and Nakanishi 2002; Kameda et al. 2011; Toyokawa et al. 2014) and is a necessary outcome where alternative tactics within a population depend on both their frequency and the phenotypes of individuals (Repka and Gross 1995; Gross 1996; Barta and Giraldeau 1998).

Reducing uncertainty in interactions can therefore be achieved by having fewer but more frequent interactions with a set of specific (known) individuals. Where individuals join and leave groups frequently and the system displays high fission–fusion dynamics (Sueur et al. 2011), uncertainty reduction can be achieved via a signal or cue indicative of a particular individual strategy (identity). For instance, in Gouldian finches, *Erythrura gouldiae*, which show two major head-color morphs (red and black), black-headed birds are bolder but less aggressive than red-headed birds (Williams et al. 2012). It has been hypothesized that these head color personality correlations may minimize competitive interactions and facilitate cooperative interactions in groups (Williams et al. 2012; King et al. 2015). Similarly, in species ranging from ants to fish to mammals, group identity can be mediated by olfactory cues (Zenuto and Fanjul 2002; Matsumura et al. 2007) and shared information transmission, including horizontal gene transfer (Ochman et al. 2000), and might be key to explaining the emergence of group identity and resulting cooperative behavior. This is because associations (and thus likelihood of interactions) formed on the basis of such "group identities" or "strategy identities" can reduce the strategic uncertainty that any individual has to face. Work by Ehrhardt and Keser (1999) illustrates how this process may work. In their experiment with human subjects, individuals can endogenously select in which group they interact to play a public goods game (i.e., a multiperson prisoner's dilemma game). They found that cooperators try to be with cooperators, but selfish persons (i.e., exploiters) are constantly chasing the cooperators, which leads to low stability in the groups and a high degree of uncertainty. In Ehrhardt and Keser's setting, cooperators cannot prevent selfish people from joining their group (Charness et al. 2014). Thus, controlling group membership or getting individuals to comply with group-specific behavioral prescriptions (e.g., social norms) can be a means of preserving cooperation within the group.

In the case of humans, who exhibit strategic behavior (i.e., take into account the strategic rationality of other players), the information structure which mediates the emergence of social norm compliance and mutual cooperation might take more complex forms. In particular, one of the main mechanisms enabling coordination on efficient outcomes in social dilemmas has been shown to be conditional cooperation (Fischbacher et al. 2001), which allows for an evolutionary explanation (Mengel 2008). In this case, a stable cooperation can be achieved only if a given individual *expects* that others will also cooperate. With respect to group identity, this requires coordination not only at the level of first-order beliefs (i.e., beliefs about whether oneself and other individuals belong to a group), but also at the level of second-order beliefs (whether others also *believe* that they belong to the group, and hence might comply with the group norms as well). Otherwise the cooperation of opposing individuals cannot be predicted in advance.

In game theoretic terms, mutual cooperation of conditional players (at least in humans) might therefore require *common knowledge* of group identity. A series of recent laboratory experiments confirmed that common knowledge might be a crucial prerequisite for the emergence of cooperation and prosocial behavior in human groups (thus, truly "activating" group identity), even with minimal group identity (Yamagishi and Mifune 2008; Guala et al. 2013; Ockenfels and Werner 2014). Complementing theoretical work, Gintis (2010) suggests that common knowledge of within-group norms might be interpreted as a purely signaling device, facilitating coordination on socially efficient equilibria and hence self-sufficient for establishing cooperation. Further research in this direction is needed to shed light on the extent to which common knowledge (or similar information structures) might explain the emergence and evolution of real-life cooperation (Henrich et al. 2010).

Another important factor of stability of cooperation within social groups is the perceived *fairness* of internal group rules and procedures; for instance, with respect to the distribution of commonly produced outcome, individual autonomy, or sanctions for defectors (Hartner et al. 2008; Kosfeld et al. 2009; Gaechter and Thöni 2010). Various models suggest several key criteria for assessing fairness in strategic interactions (Fehr and Schmidt 1999; Bolton and Ockenfels 2000; Falk and Fischbacher 2006). At the same time, empirical evidence shows that both individuals and communities strongly vary in their perception of fairness (Henrich et al. 2010; Gelfand et al. 2011). This means that constructing a universal theory of fairness, taking into account its dependence on the situational and institutional context (Henrich et al. 2010; Falk and Szech 2013), still remains an open area for research. In terms of institutional design, a related question is how to reconcile and strengthen the perception of fairness within heterogeneous groups, which might again call for coordinating information structures such as common knowledge of group identity or shared contextual framing (Ellingsen et al. 2012).

In addition to institutional design, there are both physical (Oliver 1993; King and Sueur 2011) and cognitive (Dunbar 2003; Dunbar and Shultz 2007) limits to maintaining identity in groups, and the benefits of pooling efforts also diminishes in larger group sizes (King and Cowlishaw 2007). Laboratory experiments support this because norm enforcement (e.g., through altruistic punishment of defectors) can be particularly strong in small groups (Henrich et al. 2010; Perc et al. 2013), where a deviation from the group norm is relatively more salient and, hence, more detrimental for the overall group identity. Finally, large groups might be sensitive to the emergence of opportunistic subgroups (characterized by cooperation within the subgroup) but, at the same time, exploit the total group outcome (for further discussion, see Foster et al., this volume). These seemingly contradictory behaviors (prosocial with respect to the subgroup but opportunistic with respect to the whole group) suggest further that intra-(sub)group cooperation might be largely viewed as a pragmatic method of coordination on an efficient outcome, rather than a revelation of altruistic preferences (Guala et al. 2013).

Consequences of Exploitation for Institutions: Intergroup Relations and Economic Conflict Theory

Once groups are maintained and cooperate to produce and share resources, they may become open to exploitation by other groups. Foster et al. (this volume) explore the challenges for market and institutional design when countering exploitation strategies. Here, we take a look at economic conflict theory (ECT) (Hirshleifer 1989, 1991; Skaperdas 1996), which we believe holds much promise for exploring exploitative behavior between groups. The idea behind ECT is to treat appropriative and/or defensive actions as economic activities in their own right. ECT assumes the absence of well-defined property rights in resources and goods in a state of anarchy, which makes expropriation of other owners of resources or goods (through force or otherwise) a viable alternative to own production. Hence there is a trade-off in resource use between productive and appropriative measures. This view goes beyond the scenario of producer–scrounger or hawk–dove games developed in evolutionary biology (see Dubois et al. and Burton-Chellew et al., this volume). These games allow for a dichotomous (extreme) set of actions, but ECT is very applicable to evolutionary analysis with minimal conceptual tweaking.

ECT assumes that two economic agents (which can be individuals, groups, or populations) can combine their resources, R_1 and R_2, in a productive effort to create an additional surplus (consumption good). Evolutionary biologists would refer to this as a public/communal good, and hawk–dove analysis assumes such a communally available resource exists *a priori*. The two agents then have to agree on a division of this surplus. Although their preferences are aligned with regard to the production of the additional surplus (both prefer cooperation), they are diametrically opposed with respect to its distribution (each prefers more over less). In the standard economic model with well-established property rights (which, moreover, can be enforced at no cost), rational individuals would peacefully agree on a division through some bargaining procedure, or through a contract that is specified in advance, which is assumed to be enforceable. However, in a state of anarchy without established property rights, the same rational agents would dispute the produced good once it is produced. This provides the rationale to invest part of the resources into "weapons," which increase the likelihood of success in any ensuing conflict over the produced goods. The marginal return on an additional unit of resource invested into weapons is measured by the additional expected winnings from the contest over the produced good. This is smaller now, since one less unit of resources is used in its production, but the probability of winning this smaller "pie" has increased for the investing player. The marginal cost of this additional investment is given by the individual after-fight share of the amount of the good that could have been produced; the latter depends on the productivity of the investing agent. Obviously, the first option—no one invests in weapons and all resources go toward the production of the good—is not an equilibrium.

A conflict technology (contest success function) transforms the investments in weapons made by both players into winning probabilities for both players (Hirshleifer 1989). If neither player invests, a fair division is assumed; investment by only one player leads to a winning probability of "1" for this player. When both players increase their investments into weapons, the amount of resources invested in the production of the good decreases. Symmetric players (i.e., those with identical amounts of the resource and identical productivity) will invest the same amount of resources into weapons. Hence both players win the contest over the produced good with probability $1/2$. If $R_1 > R_2$, this result still applies. Hirshleifer (1989, 1991) coined this the *paradox of power*: a poorer agent turns out to be equally powerful in the contest over the produced good. This paradox becomes more pronounced if the agents also differ in productivity for the good: the *more* productive player invests always *less* into weapons in equilibrium than the less productive one. Hence, the *less* productive agent will be *more* powerful in the contest for the produced good and will achieve a higher payoff in equilibrium. Whereas the more productive agent enjoys a comparative advantage in production, the less productive one has the advantage in fighting. This means that productiveness is *inversely* related to power, which is opposite to the case of secure property rights. Moreover, more effective weapons lead to more investments into weapons, and hence lower production and material welfare. Generalizations of this model are provided by Skaperdas (1996) and Hwang (2009).

A slightly more elaborate model allows for the peaceful settlement of the contest over the produced product (Garfinkel and Skaperdas 2006). Suppose that after both agents have invested into weapons and production they bargain over the division of the output. If an agreement is reached at this stage, both agents share the good accordingly and the game ends. If no agreement is reached, both agents fight it out with their weapons in a third stage. It is possible to reach agreement at the second stage in the shadow of the conflict that looms ahead (at the third stage). Thus investment into weapons now serves the purpose of gaining an advantageous bargaining position. Clearly, in the bargaining solution, each player must at least receive the amount he can expect from the conflict. The exact division depends on the used bargaining procedure, which also influences investments into weapons.

This model has been adapted to production and conflict between groups (see, e.g., Skaperdas 1998; Wärneryd 1998). Although it is not immediately obvious to us where ECT may apply biologically to a particular species or context, we expect that the same sort of conflict could manifest in life history trade-offs where (say) growing weaponry (or musculature) will diminish the resources available for effort into activities that can produce something that is potentially shareable (Isbell 1991; Berglund et al. 1996). On a longer term, the inverse relationship between productiveness and power, as set out by ECT, gives disincentives to productive innovations and hence stalls growth.

Conclusion

We have managed to go through this chapter with only limited reference to game theoretical models. Mengel and van der Weele, Burton-Chellew et al., and Barta (this volume) provide full details and examples of how these models can be used to understand exploitative strategies. This does not mean that we dismiss this approach; rather the opposite. Research in exploitation by both economists and biologists is related directly or indirectly to the use of game theoretical models. We believe that regular movement back and forth between these models, experimental data, and statistical fitting are necessary to begin to understand how and why exploitation occurs in different biological systems. Systematically classifying any biological system through the use of models may allow us to reverse engineer the key ingredients. Then we may be in a position to manipulate the payoffs and reduce exploitation where this impacts negatively on a system.

Acknowledgments

We wish to thank Juliette Hyde for her assistance during the Forum, Luc-Alain Giraldeau and Philipp Heeb, as chairpersons, and all of those who joined in our discussions to make this chapter possible. Sasha R. X. Dall was supported by a Leverhulme Trust International Network Grant to Sasha R. X. Dall, Peter Hammerstein, Olof Leimar, and John McNamara.

Bibliography

Acerbi, A., and A. Mesoudi. 2015. If We Are All Cultural Darwinians What's the Fuss About? Clarifying Recent Disagreements in the Field of Cultural Evolution. *Biol. Philos.* **30**:481–503. [3]

Afshar, M., and L.-A. Giraldeau. 2014. A Unified Modelling Approach for Producer-Scrounger Games in Complex Ecological Conditions. *Anim. Behav.* **96**:167–176. [3]

Afshar, M., C. L. Hall, and L.-A. Giraldeau. 2015. Zebra Finches Scrounge More When Patches Vary in Quality: Experimental Support of the Linear Operator Learning Rule. *Anim. Behav.* **105**:181–186. [3]

Agosta, S. J., N. Janz, and D. R. Brooks. 2010. How Specialists Can Be Generalists: Resolving the "Parasite Paradox" and Implications for Emerging Infectious Disease. *Zoologia* **27**:151–162. [3]

Agrawal, A. 2001. Common Property Institutions and the Sustainable Governance of Resources. *World Dev.* **29**:1649–1672. [5]

Agrawal, A., and A. Chhatre. 2006. Explaining Success on the Commons: Community Forest Governance in the Indian Himalayas. *World Dev.* **34**:149–166. [5]

Agrawal, A., A. Chhatre, and R. Hardin. 2008. Changing Governance of the World's Forests. *Science* **320**:1460–1462. [5]

Agrawal, A., and S. Goyal. 2001. Group Size and Collective Action: Third-Party Monitoring in Common-Pool Resources. *Comp. Polit. Stud.* **34**:63–93. [5]

Agrawal, A., and E. Ostrom. 2001. Collective Action, Property Rights, and Decentralization in Resource Use in India and Nepal. *Polit. Soc.* **29**:485–514. [5]

Agrawal, A., and G. Yadama. 1997. How Do Local Institutions Mediate Market and Population Pressures on Resources? Forest Panchayats in Kumaon, India. *Dev. Change* **28**:435–465. [5]

Akerlof, G. A., and R. E. Kranton. 2005. Identity and the Economics of Organizations. *J. Econ. Perspect.* **19**:9–32. [10]

Alger, I. and J. W. Weibull. 2013. Homo Moralis: Preference Evolution under Incomplete Information and Assortative Matching. *Econometrica* **81**:2269–2302. [2]

Allen, B., and M. A. Nowak. 2015. Games among Relatives Revisited. *J. Theor. Biol.* **378**:103–116. [3]

Allen, R. C., L. McNally, R. Popat, and S. P. Brown. 2016. Quorum Sensing Protects Bacterial Cooperation from Exploitation by Cheats. *ISME J.* **10**:1706–1716. [7, 10]

Allen, R. C., R. Popat, S. P. Diggle, and S. P. Brown. 2014. Targeting Virulence: Can We Make Evolution-Proof Drugs? *Nat. Rev. Microbiol.* **12**:300–308. [7]

Allingham, M. 2002. Choice Theory: A Very Short Introduction. Oxford: Oxford Univ. Press. [3]

Allport, G. 1954. The Nature of Prejudice. Reading, MA: Addison-Wesley. [2]

Alzua, M. L., J. C. Cardenas, and H. Djebbari. 2014. Community Mobilization around Social Dilemmas: Evidence from Lab Experiments in Rural Mali. *CEDLAS Working Papers* **160**:1–34. [6]

Ambrus, A., and B. Greiner. 2012. Imperfect Public Monitoring with Costly Punishment: An Experimental Study. *Am. Econ. Rev.* **102**:3317–3332. [10]

Ambrus, A., B. Greiner, and P. Pathak. 2015. How Individual Preferences Are Aggregated in Groups: An Experimental Study. *J. Public Econ.* **129**:1–13. [10]

Andam, K. S., P. J. Ferraro, K. R. E. Sims, A. Healy, and M. B. Holland. 2010. Protected Areas Reduced Poverty in Costa Rica and Thailand. *PNAS* **107**:9996–10001. [5]

Andersson, M. 1985. Brood Parasites within Species. In: Producers and Scroungers, ed. C. J. Barnard, pp. 195–228. New York: Chapman & Hall. [11]

Andersson, M., and J. Krebs. 1978. Evolution of Hoarding Behavior. *Anim. Behav.* **26**:707–711. [12]

Andreoni, J. 1995. Cooperation in Public-Goods Experiments: Kindness or Confusion. *Am. Econ. Rev.* **85**:891–904. [2, 3]

Andreoni, J., and J. H. Miller. 1993. Rational Cooperation in the Finitely Repeated Prisoners Dilemma: Experimental Evidence. *Econ. J.* **103**:570–585. [2, 3]

Aoki, K., L. Lehmann, and M. W. Feldman. 2011. Rates of Cultural Change and Patterns of Cultural Accumulation in Stochastic Models of Social Transmission. *Theor. Popul. Biol.* **79**:192–202. [3]

Arbilly, M. 2015. Understanding the Evolution of Learning by Explicitly Modeling Learning Mechanisms. *Curr. Zool.* **61**:341–349. [3]

Arbilly, M., U. Motro, M. W. Feldman, and A. Lotem. 2010. Co-Evolution of Learning Complexity and Social Foraging Strategies. *J. Theor. Biol.* **267**:573–581. [3]

Arbilly, M., D. B. Weissman, M. W. Feldman, and U. Grodzinski. 2014. An Arms Race between Producers and Scroungers Can Drive the Evolution of Social Cognition. *Behav. Ecol.* **25**:487–495. [3]

Archetti, M., and I. Scheuring. 2011. Coexistence of Cooperation and Defection in Public Goods Games. *Evolution* **65**:1140–1148. [3]

———. 2012. Review: Game Theory of Public Goods in One-Shot Social Dilemmas without Assortment. *J. Theor. Biol.* **299**:9–20. [6]

Aumann, R. J. 1995. Backward Induction and Common Knowledge of Rationality. *Games Econ. Behav.* **8**:6–19. [3]

Aureli, F., and C. P. Vanschaik. 1991. Post-Conflict Behavior in Long-Tailed Macaques (*Macaca fascicularis*): Coping with the Uncertainty. *Ethology* **89**:101–114. [12]

Avitabile, A., M. Herold, G. B. M. Heuvelink, et al. 2016. An Integrated Pan-Tropical Biomass Map Using Multiple Reference Datasets. *Global Change Biol.* **22**:1406–1420. [5]

Axelrod, R. 2000. On Six Advances in Cooperation Theory. *Analyse & Kritik* **22**:130–151. [3]

Axelrod, R., and W. D. Hamilton. 1981. The Evolution of Cooperation. *Science* **211**:1390–1396. [3, 6]

Baland, J. M., P. Bardhan, and S. Bowles, eds. 2006. Inequality, Cooperation, and Environmental Sustainability. Princeton: Princeton Univ. Press. [6]

Balooni, K., M. Inoue, and V. Ballabh. 2007. Declining Instituted Collective Management Practices and Forest Quality in the Central Himalayas. *Econ. Polit. Wkly.* **42**:1443–1452. [5]

Baquero, F., and J. Campos. 2003. The Tragedy of the Commons in Antimicrobial Chemotherapy. *Rev. Esp. Quimioter.* **16**:11–13. [9]

Bardhan, P., and D. Mookherjee. 2002. Relative Capture of Local and Central Governments: An Essay in the Political Economy of Decentralization. Secondary Relative Capture of Local and Central Governments: An Essay in the Political Economy of Decentralization. http://escholarship.org/uc/item/9gx7t5hd. (accessed Aug. 18, 2016). [6]

Barnard, C. J., ed. 1984. Producers and Scroungers: Strategies of Exploitation and Parasitism. London: Croom Helm. [3]

Barnard, C. J., and J. M. Behnke, eds. 1990. Parasitism and Host Behaviour. London: Taylor & Francis. [3]

Barnard, C. J., and R. M. Sibly. 1981. Producers and Scroungers: A General-Model and Its Application to Captive Flocks of House Sparrows. *Anim. Behav.* **29**:543–550. [3, 6, 11, 12]

Barrette, M., and L.-A. Giraldeau. 2006. Prey Crypticity Reduces the Proportion of Group Members Searching for Food. *Anim. Behav.* **71**:1183–1189. [11]

Barta, Z., R. Flynn, and L.-A. Giraldeau. 1997. Geometry for a Selfish Foraging Group: A Genetic Algorithm Approach. *Proc. R. Soc. Lond. B* **264**:1233–1238. [3, 11]

Barta, Z., and L.-A. Giraldeau. 1998. The Effect of Dominance Hierarchy on the Use of Alternative Foraging Tactics: A Phenotype-Limited Producing-Scrounging Game. *Behav. Ecol. Sociobiol.* **42**:217–223. [4, 12]

———. 2000. Daily Patterns of Optimal Producer and Scrounger Use under Predation Hazard: A State-Dependent Dynamic Game Analysis. *Am. Nat.* **155**:570–582. [3, 4]

———. 2001. Breeding Colonies as Information Centres: A Re-Appraisal of Information-Based Hypotheses Using the Producer-Scrounger Game. *Behav. Ecol.* **12**:121–127. [3, 4]

Baselski, V. S., R. A. Medina, and C. D. Parker. 1978. Survival and Multiplication of *Vibrio cholerae* in the Upper Bowel of Infant Mice. *Infect. Immun.* **22**:435–440. [9]

———. 1979. *In Vivo* and *in Vitro* Characterization of Virulence-Deficient Mutants of *Vibrio cholerae*. *Infect. Immun.* **24**:111–116. [9]

Basurto, X., E. Blanco, M. Nenadovic, and B. Vollan. 2016. Integrating Simultaneous Prosocial and Antisocial Behavior into Theories of Collective Action. *Science Advances* **2**:e1501220. [3]

Batisse, M. 1997. Biosphere Reserves: A Challenge for Biodiversity Conservation and Regional Development. *Environment* **39**:6–13. [5]

Bauch, C. T., and D. J. D. Earn. 2004. Vaccination and the Theory of Games. *PNAS* **101**:13391–13394. [9]

Baylis, K., J. Honey-Rosés, J. Börner, et al. 2016. Mainstreaming Impact Evaluation in Nature Conservation. *Conserv. Lett.* **9**:58–64. [5]

Baynes, J., J. Herbohn, C. Smith, R. Fisher, and D. Bray. 2015. Key Factors Which Influence the Success of Community Forestry in Developing Countries. *Global Environ. Change* **35**:226–238. [5]

Beauchamp, G. 2000. Learning Rules for Social Foragers: Implications for the Producer-Scrounger Game and Ideal Free Distribution Theory. *J. Theor. Biol.* **207**:21–35. [3]

———. 2008. A Spatial Model of Producing and Scrounging. *Anim. Behav.* **76**:1935–1942. [3]

Becker, D., M. Selbach, C. Rollenhage, et al. 2006. Robust Salmonella Metabolism Limits Possibilities for New Microbials. *Nature* **440**:303–307. [8]

Becker, G. 1968. Crime and Punishment: An Economic Approach. *J. Polit. Econ.* **76**:169–217. [2]

Belmaker, A., U. Motro, M. W. Feldman, and A. Lotem. 2012. Learning to Choose among Social Foraging Strategies in Adult House Sparrows (*Passer domesticus*). *Ethology* **118**:1111–1121. [3]

Belshaw, C. S. 1965. The Cultural Milieu of the Entrepreneur. In: Explorations in Enterprise, ed. H. G. Aitken, pp. 139–162. Cambridge, MA: Harvard Univ. Press. [3]

Bénabou, R., and J. Tirole. 2006. Incentives and Prosocial Behavior. *Am. Econ. Rev.* **96**:1652–1677. [2]

Berglund, A., A. Bisazza, and A. Pilastro. 1996. Armaments and Ornaments: An Evolutionary Explanation of Traits of Dual Utility. *Biol. J. Linn. Soc.* **58**:385–399. [12]

Bergmüller, R., and M. Taborsky. 2010. Animal Personality Due to Social Niche Specialisation. *Trends Ecol. Evol.* **25**:504–511. [11]

Bergstrom, C. T., and P. Godfrey-Smith. 1998. On the Evolution of Behavioral Heterogeneity in Individuals and Populations. *Biol. Philos.* **13**:205–231. [11]

Bester, H., and W. Gueth. 1998. Is Altruism Evolutionarily Stable? *J. Econ. Behav. Organ.* **34**:193–209. [2]

Bikhchandani, S., D. Hirshleifer, and I. Welch. 1998. Learning from the Behavior of Others: Conformity Fads and Informational Cascades. *J. Econ. Perspect.* **12**:151–170. [11]

Binmore, K. 1996. A Note on Backward Induction. *Games Econ. Behav.* **17**:135–137. [3]

Binmore, K., J. McCarthy, G. Ponti, L. Samuelson, and A. Shaked. 2002. A Backward Induction Experiment. *J. Econ. Theory* **104**:48–88. [3]

Bisin, A., G. Topa, and T. Verdier. 2004. Cooperation as a Transmitted Cultural Trait. *Ration. Soc.* **16**:477–507. [2]

Blanco, M., D. Engelmann, and H. T. Normann. 2011. A Within-Subject Analysis of Other-Regarding Preferences. *Games Econ. Behav.* **72**:321–338. [2]

Bloch, M. 1973. The Long Term and the Short Term: The Economic and Political Significance of the Morality of Kinship. In: The Character of Kinship, ed. J. Goody, pp. 75–87. London: Cambridge Univ. Press. [3]

Bohnet, I., B. Frey, and S. Huck. 2001. More Order with Less Law: On Contract Enforcement, Trust, and Crowding. *Am. Polit. Sci. Rev.* **95**:131–144. [2]

Bolton, G., B. Greiner, and A. Ockenfels. 2013. Engineering Trust: Reciprocity in the Production of Reputation Information. *Manag. Sci.* **59**:265–285. [10]

———. 2015. Conflict Resolution vs. Conflict Escalation in Online Markets. Secondary Conflict Resolution vs. Conflict Escalation in Online Markets. http://research.economics.unsw.edu.au/RePEc/papers/2015-19.pdf. (accessed Aug. 25, 2016). [10]

Bolton, G. E., E. Katok, and A. Ockenfels. 2004. How Effective Are Electronic Reputation Mechanisms? An Experimental Investigation. *Manag. Sci.* **50**:1587–1602. [2]

Bolton, G. E., and A. Ockenfels. 2000. ERC: A Theory of Equity, Reciprocity, and Competition. *Am. Econ. Rev.* **90**:166–193. [3, 12]

Bonhoeffer, S., M. Lipsitch, and B. R. Levin. 1997. Evaluating Treatment Protocols to Prevent Antibiotic Resistance. *PNAS* **94**:12106–12111. [8, 9]

Börgers, T., and R. Sarin. 1997. Learning through Reinforcement and Replicator Dynamics. *J. Econ. Theory* **77**:1–14. [2, 3]

Börner, J., and S. Vosti. 2013. Managing Tropical Forest Ecosystem Services: An Overview of Options. In: Governing the Provision of Ecosystem Services, ed. R. Muradian and L. Rival, pp. 21–46. Dordrecht: Springer. [6]

Bowler, D. E., L. M. Buyung-Ali, J. R. Healy, et al. 2012. Does Community Forest Management Provide Global Environmental Benefits and Improve Local Welfare? *Front. Ecol. Environ.* **10**:29–36. [5]

Bowles, S. 1998. Endogenous Preferences: The Cultural Consequences of Markets and Other Economic Institutions. *J. Econ. Lit.* **36**:75–111. [2]

Bowles, S., and S. H. Hwang. 2008. Social Preferences and Public Economics: Mechanism Design When Social Preferences Depend on Incentives. *J. Public Econ.* **92**:1811–1820. [3]

Bowles, S., and S. Polania-Reyes. 2012. Economic Incentives and Social Preferences: Substitutes or Complements? *J. Econ. Lit.* **50**:368–425. [2, 6]

Boyd, R., and P. J. Richerson. 1995. Why Does Culture Increase Human Adaptability. *Ethol. Sociobiol.* **16**:125–143. [3]

———. 2005. The Origin and Evolution of Cultures. Evolution and Cognition, S. Stich, series ed. Oxford: Oxford Univ. Press. [2]

Branas-Garza, P., R. Cobo-Reyes, M. P. Espinosa, et al. 2010. Altruism and Social Integration. *Games Econ. Behav.* **69**:249–257. [2]

Bray, D. B., L. Merino-Pérez, P. Negreros-Castillo, et al. 2003. Mexico's Community-Managed Forests as a Global Model for Sustainable Landscapes. *Conserv. Biol.* **17**:672–677. [5]

Brown, S. P., D. M. Cornforth, and N. Mideo. 2012. Evolution of Virulence in Opportunistic Pathogens: Generalism, Plasticity, and Control. *Trends Microbiol.* **20**:336–342. [7, 9]

Brown, S. P., R. F. Inglis, and F. Taddei. 2009a. Evolutionary Ecology of Microbial Wars: Within-Host Competition and (Incidental) Virulence. *Evol. Appl.* **2**:32–39. [4, 7]

Brown, S. P., and P. D. Taylor. 2010. Joint Evolution of Multiple Social Traits: A Kin Selection Analysis. *Proc. Roy. Soc. B* **277**:415–427. [7]

Brown, S. P., S. A. West, S. P. Diggle, and A. S. Griffin. 2009b. Social Evolution in Micro-Organisms and a Trojan Horse Approach to Medical Intervention Strategies. *Phil. Trans. R. Soc. B* **364**:3157–3168. [7]

Bshary, A., and R. Bshary. 2010. Self-Serving Punishment of a Common Enemy Creates a Public Good in Reef Fishes. *Curr. Biol.* **20**:2032–2035. [6]

Buchanan, J. M. 1975. The Samaritan's Dilemma. In: Altruism, Morality and Economic Theory, ed. E. S. Phelps, pp. 71–85. New York: Russel Sage Foundation. [3]

Buckling, A., and M. A. Brockhurst. 2008. Kin Selection and the Evolution of Virulence. *Heredity* **100**:484–488. [9]

Bugnyar, T., and B. Heinrich. 2005. Ravens, *Corvus corax*, Differentiate between Knowledgeable and Ignorant Competitors. *Proc. R. Soc. Lond. B* **272**:1641–1646. [11]

———. 2006. Pilfering Ravens, *Corvus corax*, Adjust Their Behaviour to Social Context and Identity of Competitors. *Anim. Cogn.* **9**:369–376. [11]

Bugnyar, T., and K. Kotrschal. 2002. Observational Learning and the Raiding of Food Caches in Ravens, *Corvus corax*: Is It "Tactical" Deception? *Anim. Behav.* **64**:185–195. [11]

Camerer, C. 2003. Behavioral Game Theory: Experiments in Strategic Interaction. Princeton: Princeton Univ. Press. [2]

Camerer, C., and T. H. Ho. 1999. Experience-Weighted Attraction Learning in Normal Form Games. *Econometrica* **67**:827–874. [3]

Caraco, T., and L.-A. Giraldeau. 1991. Social Foraging: Producing and Scrounging in a Stochastic Environment. *J. Theor. Biol.* **153**:559–583. [3, 11]

Cavalli-Sforza, L. L., and M. Feldman. 1981. Cultural Transmission and Evolution: A Quantitative Approach. Princeton: Princeton Univ. Press. [2]

Chadwick, D. J., and J. Goode, eds. 1997. Antibiotic Resistance: Origins, Evolution, Selection, and Spread. CIBA Foundation Symposium 207. New York: Wiley. [9]

Charness, G., R. Cobo-Reyes, and N. Jimenez. 2014. Identities, Selection, and Contributions in a Public-Goods Game. *Games Econ. Behav.* **87**:322–338. [12]

Charness, G., and M. Duwfenberg. 2006. Promises and Partnerships. *Econometrica* **74**:1579–1601. [2]

Charness, G., and M. Rabin. 2002. Understanding Social Preferences with Simple Tests. *Q. J. Econ.* **117**:817–869. [3, 4]

Chaudhuri, A. 2011. Sustaining Cooperation in Laboratory Public Goods Experiments: A Selective Survey of the Literature. *Exp. Econ.* **14**:47–83. [3, 10]

Cheung, Y. W., and D. Friedman. 1997. Individual Learning in Normal Form Games: Some Laboratory Results. *Games Econ. Behav.* **19**:46–76. [3]

Chhatre, A., and A. Agrawal. 2008. Forest Commons and Local Enforcement. *PNAS* **105**:13286–13291. [5]

Chichilnisky, G. 1996. An Axiomatic Approach to Sustainable Development. *Soc. Choice Welfare* **13**:231–257. [9]

Choffnes, E. R., D. A. Relman, and A. Mack. 2010. Antibiotic Resistance: Implications for Global Health and Novel Intervention Strategies: Workshop Summary. Washington, D.C.: National Academies Press. [9]

Christakis, N., and J. Fowler. 2008. The Collective Dynamics of Smoking in a Large Social Network. *New Engl. J. Med.* **358**:2249–2259. [2]

Christian, C., D. Ainley, M. Bailey, et al. 2013. A Review of Formal Objections to Marine Stewardship Council Fisheries Certifications. *Biol. Conserv.* **161**:10–17. [6]

Clark, C. W. 1990. Mathematical Bioeconomics: The Optimal Management of Renewable Resources (2nd edition). New York: Wiley. [6]

Clark, C. W., and M. Mangel. 1986. The Evolutionary Advantages of Group Foraging. *Theor. Popul. Biol.* **3**:45–75. [6]

Clatworthy, A. E., E. Pierson, and D. T. Hung. 2007. Targeting Virulence: A New Paradigm for Antimicrobial Therapy. *Nat. Chem. Biol.* **3**:541–548. [7]

Clayton, N. C., D. P. Griffiths, N. D. Emery, and A. Dickinson. 2001. Elements of Episodic-Like Memory in Animals. *Phil. Trans. R. Soc. B* **356**:1483–1491. [11]

Clutton-Brock, T. H., and G. A. Parker. 1995. Punishment in Animal Societies. *Nature* **373**:209–216. [12]

Coates, A. R. M., G. Halls, and Y. Hu. 2011. Novel Classes of Antibiotics or More of the Same? *Br. J. Pharmacol.* **163**:184–194. [8]

Congleton, R. D., A. L. Hillman, and K. A. Konrad. 2008. 40 Years of Research on Rent Seeking: Theory of Rent Seeking. Berlin: Springer. [3]

Coolen, I. 2002. Increasing Foraging Group Size Increases Scrounger Use and Reduces Searching Efficiency in Nutmeg Mannikins (*Lonchura punctulata*). *Behav. Ecol. Sociobiol.* **52**:232–238. [3, 11]

Coolen, I., and L.-A. Giraldeau. 2003. Incompatibility between Anti-Predatory Vigilance and Scrounger Tactic in Nutmeg Mannikins (*Lonchura punctulata*). *Anim. Behav.* **66**:657–664. [4]

Coolen, I., L.-A. Giraldeau, and M. Lavoie. 2001. Head Position as an Indicator of Producer and Scrounger Tactics in a Ground Feeding Bird. *Anim. Behav.* **61**:895–903. [3, 4, 11]

Coolen, I., L.-A. Giraldeau, and W. L. Vickery. 2007. Scrounging Behavior Regulates Population Dynamics. *Oikos* **116**:533–539. [4, 11]

Coolen, I., A. J. W. Ward, P. J. B. Hart, and K. N. Laland. 2005. Foraging Nine-Spined Sticklebacks Prefer to Rely on Public Information over Simpler Social Cues. *Behav. Ecol.* **16**:865–870. [11]

Cote, I. M., and R. Poulin. 1995. Parasitism and Group-Size in Social Animals: A Metaanalysis. *Behav. Ecol.* **6**:159–165. [12]

Courvalin, P. 2010. Antibiotic-Induced Resistance Flow. In: Antibiotic Resistance: Implications for Global Health and Novel Intervention Strategies: Workshop Summary, ed. E. R. Choffnes et al., pp. 141–149. Washington, D.C.: National Academies Press. [9]

Cox, M., G. Arnold, and T. S. Villamayor. 2010. A Review of Design Principles for Community-Based Natural Resource Management. *Ecol. Soc.* **15**:38. [5]

Crawford, S. E. S., and E. Ostrom. 1995. A Grammar of Institutions. *Am. Polit. Sci. Rev.* **89**:582–600. [6]

Crespi, B. J. 2001. The Evolution of Social Behavior in Microorganisms. *Trends Ecol. Evol.* **16**:178–183. [7]

Croft, D. P., J. Krause, S. K. Darden, et al. 2009. Behavioural Trait Assortment in a Social Network: Patterns and Implications. *Behav. Ecol. Sociobiol.* **63**:1495–1503. [12]

Croson, R., and M. Marks. 2000. Step Returns in Threshold Public Goods: A Meta- and Experimental Analysis. *Exp. Econ.* **2**:239–259. [3]

Dall, S. R. X., A. I. Houston, and J. M. M. McNamara. 2004. The Behavioural Ecology of Personality: Consistent Individual Differences from an Adaptive Perspective. *Ecol. Lett.* **7**:734–739. [11]

Dally, J. M., N. J. Emery, and N. S. Clayton. 2005. Cache Protection Strategies by Western Scrub-Jays *Aphelocoma californica*: Implications for Social Cognition. *Anim. Behav.* **70**:1251–1263. [11]

Danchin, E., L.-A. Giraldeau, T. J. Valone, and R. H. Wagner. 2004. Public Information: From Nosy Neighbors to Cultural Evolution. *Science* 305:487–491. [11]

Darimont, C. T. 2015. The Unique Ecology of Human Predators. *Science* 349:858. [6]

Darley, J. M., and B. Latane. 1968. Bystander Intervention in Emergencies: Diffusion of Responsibility. *J. Pers. Soc. Psychol.* **8**:377–383. [3]

Daugherty, M. D., and H. S. Malik. 2012. Rules of Engagement: Molecular Insights from Host-Virus Arms Races. *Annu. Rev. Genet.* **46**:677–700. [12]

Davies, J. 1994. New Pathogens and Old Resistance Genes. *Microbiologia* **10**:9–12. [9]

Davies, N. B. 2000. Cuckoos, Cowbirds and Other Cheats, Illustrated by David Quinn. London: T. & A. D. Poyser Ltd. [10]

Davies, N. B., J. R. Krebs, and S. A. West. 2012. An Introduction to Behavioural Ecology. Oxford: Wiley-Blackwell. [3]

Davis, S. H., and A. Wali. 1994. Indigenous Land Tenure and Tropical Forest Management in Latin America. *Ambio* **23**:485–490. [5]

Dawkins, R., and J. R. Krebs. 1979. Arms Races between and within Species. *Proc. Roy. Soc. B* **205**:489–511. [3]

D'Costa, V., C. King, L. Kalan, et al. 2011. Antibiotic Resistance Is Ancient. *Nature* **477**:457–461. [9]

Dechenaux, E., D. Kovenock, and R. M. Sheremeta. 2015. A Survey of Experimental Research on Contests, All-Pay Auctions and Tournaments. *Exp. Econ.* **18**:609–669. [3]

Dhami, S. 2016. Foundations of Behavioral Economic Analysis. Oxford: Oxford Univ. Press. [2]

Diamond, J. 1989. Overview of Recent Extinctions. In: Conservation for the Twenty First Century, ed. D. Western and M. C. Pearl, pp. 37–41. New York: Oxford Univ. Press. [6]

Díaz-Muñoz, S. L. 2011. Paternity and Relatedness in a Polyandrous Nonhuman Primate: Testing Adaptive Hypotheses of Male Reproductive Cooperation. *Anim. Behav.* **82**:563–571. [10]

Diekmann, A. 1985. Volunteers Dilemma. *J. Conflict Resolut.* **29**:605–610. [3]

———. 1993. Cooperation in an Asymmetric Volunteers Dilemma Game: Theory and Experimental-Evidence. *Int. J. Game Theory* **22**:75–85. [3]

di Falco, S., and E. Bulte. 2011. A Dark Side of Social Capital? Kinship, Consumption, and Savings. *J. Dev. Stud.* **47**:1128–1151. [3]

Diggle, S. P., A. S. Griffin, G. S. Campbell, and S. A. West. 2007. Cooperation and Conflict in Quorum-Sensing Bacterial Populations. *Nature* **450**:411–414. [7]

Dimitriu, T., C. Lotton, J. Bénard-Capelle, et al. 2014. Genetic Information Transfer Promotes Cooperation in Bacteria. *PNAS* **111**:11103–11108. [7, 10]

Dohmen, T., A. Falk, D. Huffman, and U. Sunde. 2012. The Intergenerational Transmission of Risk Attitudes. *Rev. Econ. Stud.* **79**:645–677. [2]

Dong, H., Q. Xiang, Y. Gu, et al. 2014. Structural Basis for Outer Membrane Lipopolysaccharide Insertion. *Nature* **511**:52–56. [8]

dos Santos, M., D. J. Rankin, and C. Wedekind. 2011. The Evolution of Punishment through Reputation. *Proc. R. Soc. Lond. B* **278**:371–377. [10]

Dubois, F., and L.-A. Giraldeau. 2003. The Forager's Dilemma: Food Sharing and Food Defense as Risk-Sensitive Foraging Options. *Am. Nat.* **162**:768–779. [11]

———. 2004. Reduced Resource Defence in the Absence of Information: An Experimental Test Using Captive Nutmeg Mannikins. *Anim. Behav.* **68**:21–25. [11]

———. 2005. Fighting for Resources: The Economics of Defense and Appropriation. *Ecology* **86**:3–11. [4, 11]

———. 2007. Food Sharing among Retaliators: Sequential Arrivals and Information Asymmetries. *Behav. Ecol. Sociobiol.* **62**:263–271. [11]

Dubois, F., L.-A. Giraldeau, and J. W. A. Grant. 2003. Resource Defense in a Group-Foraging Context. *Behav. Ecol.* **14**:2–9. [11]

Dubois, F., L.-A. Giraldeau, I. M. Hamilton, J. W. A. Grant, and L. Lefebvre. 2004. Distraction Sneakers Decrease the Expected Level of Aggression within Groups: A Game-Theoretic Model. *Am. Nat.* **164**:E32–E45. [11]

Dubois, F., J. Morand-Ferron, and L.-A. Giraldeau. 2010. Learning in a Game Context: Strategy Choice by Some Keeps Learning from Evolving in Others. *Proc. R. Soc. Lond. B* **277**:3609–3616. [3, 11]

Dugatkin, L. A. 1998. A Model of Coalition Formation in Animals. *Proc. R. Soc. Lond. B* **265**:2121–2125. [10]

Dunbar, R. I. M. 2003. The Social Brain: Mind, Language, and Society in Evolutionary Perspective. *Annu. Rev. Anthropol.* **32**:163–181. [12]

Dunbar, R. I. M., and S. Shultz. 2007. Evolution in the Social Brain. *Science* **317**:1344–1347. [12]

Durante, R., L. Putterman, and J. van der Weele. 2014. Preferences for Redistribution and Perception of Fairness: An Experimental Study. *J. Eur. Econ. Assoc.* **12**:1059–1086. [3]

Edmunds, D., and E. Wollenberg. 2003. Local Forest Management: The Impact of Devolution Policies. London: Earthscan. [5]

Ehrhart, K. M., and C. Keser. 1999. Mobility and Cooperation: on the Run. Working Paper 99s-24. CIRANO Série Scientifique, http://http://www.cirano.qc.ca/files/publications/99s-24.pdf. (accessed Jan. 04, 2017). [12]

Elgar, M. A. 1986. House Sparrows Establish Foraging Flocks by Giving Chirrup Calls If the Resources Are Divisible. *Anim. Behav.* **34**:169–174. [11]

Ellingsen, T., M. Johannesson, J. Mollerstrom, and S. Munkhammar. 2012. Social Framing Effects: Preferences or Beliefs? *Games Econ. Behav.* **76**:117–130. [12]

Elliot, M. 1978. Social Behavior and Foraging Ecology of the Eastern Chipmunk (*Tamias striatus*) in the Adirondack Mountains. *Smithson. Contrib. Zool.* **265**:1–107. [11]

El Mouden, C., J. B. Andre, O. Morin, and D. Nettle. 2014. Cultural Transmission and the Evolution of Human Behaviour: A General Approach Based on the Price Equation. *J. Evol. Biol.* **27**:231–241. [3]

Elster, J. 1989. The Cement of Society. Cambridge: Cambridge Univ. Press. [6]

Embrey, M., G. R. Frechette, and S. Yuksel. 2016. Cooperation in the Finitely Repeated Prisoner's Dilemma, Working Paper. *Social Science Research Network*, in press. [3]

Engel, C. 2011. Dictator Games: A Meta Study. *Exp. Econ.* **14**:583–610. [3]

Eshel, I., L. Samuelson, and A. Shaked. 1998. Altruists, Egoists and Hooligans in a Local Interaction Model. *Am. Econ. Rev.* **88**:157–179. [2]

Ewald, P. W. 1994. Evolution of Infectious Disease. New York: Oxford Univ. Press. [9]

———. 2002. Virulence Management in Humans. In: Adaptive Dynamics of Infectious Diseases: In Pursuit of Virulence Management, ed. U. Dieckmann et al., pp. 399–412. Cambridge: Cambridge Univ. Press. [9]

Falk, A., and U. Fischbacher. 2006. A Theory of Reciprocity. *Games Econ. Behav.* **54**:293–315. [12]

Falk, A., and M. Kosfeld. 2006. The Hidden Costs of Control. *Am. Econ. Rev.* **96**:1611–1630. [10]

Falk, A., and N. Szech. 2013. Morals and Markets. *Science* **340**:707–711. [12]

Farine, D. R., P. O. Montiglio, and O. Spiegel. 2015. From Individuals to Groups and Back: The Evolutionary Implications of Group Phenotypic Composition. *Trends Ecol. Evol.* **30**:609–621. [12]

Fehr, E., and S. Gaechter. 2000. Cooperation and Punishment in Public Goods Experiments. *Am. Econ. Rev.* **90**:980–994. [2, 3, 10]

———. 2002. Altruistic Punishment in Humans. *Nature* **415**:137–140. [10]

Fehr, E., and K. M. Schmidt. 1999. A Theory of Fairness, Competition, and Cooperation. *Q. J. Econ.* **114**:817–868. [3, 6, 12]

Fenichel, E. P. 2013. Economic Considerations for Social Distancing and Behavioral Based Policies During an Epidemic. *J. Health Econ.* **32**:440–451. [8]

Ferri, M., E. Ranucci, P. Romagnoli, and V. Giaccone. 2015. Antimicrobial Resistance: A Global Emerging Threat to Public Health Systems. *Crit. Rev. Food Sci. Nutr.* **13**:1040–8398. [9]

Fischbacher, U., and S. Gaechter. 2010. Social Preferences, Beliefs, and the Dynamics of Free Riding in Public Goods Experiments. *Am. Econ. Rev.* **100**:541–556. [3]

Fischbacher, U., S. Gaechter, and E. Fehr. 2001. Are People Conditionally Cooperative? Evidence from a Public Goods Experiment. *Econ. Lett.* **71**:397–404. [2, 3, 11, 12]

Flack, J. C., M. Girvan, F. B. M. de Waal, and D. C. Krakauer. 2006. Policing Stabilizes Construction of Social Niches in Primates. *Nature* **439**:426–429. [12]

Floyd, J., and D. Pauly. 1984. Smaller Size Tuna around the Philippines: Can Fish Aggregating Devices Be Blamed? *Infofish Marketing Digest* **5**:25–27. [6]

Flynn, R., and L.-A. Giraldeau. 2001. Producer-Scrounger Games in a Spatially Explicit World: Tactic Use Influences Flock Geometry of Spice Finches. *Ethology* **107**:249–257. [3, 11]

Fosco, C., and F. Mengel. 2011. Cooperation through Imitation and Exclusion in Networks. *J. Econ. Dyn. Control* **35**:641–658. [2]

Foster, D. P., and P. H. Young. 2006. Regret Testing: Learning to Play Nash Equilibrium without Knowing You Have an Opponent. *Theor. Econ.* **1**:341–367. [3]

Foster, G., and P. Frijters. 2016. Behavioral Political Economy. In: Routledge Handbook of Behavioral Economics, ed. R. Frantz et al., pp. 348–364. London: Routledge. [10]

Foster, K. R., T. Wenseleers, and F. L. W. Ratnieks. 2001. Spite: Hamilton's Unproven Theory. *Ann. Zool. Fenn.* **38**:229–238. [12]

Frank, S. A. 2003. Repression of Competition and the Evolution of Cooperation. *Evolution* **57**:693–705. [4]

Franz, M., and C. L. Nunn. 2009. Network-Based Diffusion Analysis: A New Method for Detecting Social Learning. *Proc. R. Soc. Lond. B* **276**:1829–1836. [12]

Frijters, P., and G. Foster. 2013. An Economic Theory of Greed, Love, Groups, and Networks. Cambridge: Cambridge Univ. Press. [10]

Fudenberg, D., and D. K. Levine. 1998. The Theory of Learning in Games. Cambridge, MA: MIT Press. [3]

Fürtbauer, I., A. Pond, M. Heistermann, and A. J. King. 2015. Personality, Plasticity and Predation: Linking Endocrine and Behavioural Reaction Norms in Stickleback Fish. *Funct. Ecol.* **29**:931–940. [12]

Gaechter, S. 2007. Conditional Cooperation: Behavioural Regularities from the Lab and the Field and Their Policy Implications. In: Psychology and Economics: A Promising New Cross-Disciplinary Field, ed. B. S. Frey and A. Stuzter, pp. 19–50. Cambridge, MA: MIT Press. [3]

Gaechter, S., and E. Renner. 2014. Leaders as Role Models for the Voluntary Provision of Public Goods. Secondary Leaders as Role Models for the Voluntary Provision of Public Goods. https://www.cesifo-group.de/ifoHome/publica-tions/working-papers/CESifoWP/CESifoWPdetails?wp_id=19126664. (accessed Aug. 18, 2016). [6]

Gaechter, S., E. Renner, and M. Sefton. 2008. The Long-Run Benefits of Punishment. *Science* **322**:1510. [2]

Gaechter, S., and C. Thöni. 2010. Social Comparison and Performance: Experimental Evidence on the Fair Wage–Effort Hypothesis. *J. Econ. Behav. Organ.* **76**:531–543. [12]

Gardner, A., S. A. West, and G. Wild. 2011. The Genetical Theory of Kin Selection. *J. Evol. Biol.* **24**:1020–1043. [3, 7]

Garfinkel, M. R., and S. Skaperdas. 2006. Economics of Conflict: An Overview. University of California-Irvine, Dept. of Economics, Working Paper Nr. 050623. Irvine: Univ. of California. [12]

Gautam, A. P., and A. G. Shivakoti. 2005. Conditions for Successful Local Collective Action in Forestry: Some Evidence from the Hills of Nepal. *Soc. Nat. Resour.* **18**:153–171. [5]

Geist, H. J., and E. F. Lambin. 2002. Proximate Causes and Underlying Driving Forces of Tropical Deforestation. *Bioscience* **52**:143–150. [5]

Gelfand, M. J., J. L. Raver, L. Nishii, et al. 2011. Differences between Tight and Loose Cultures: A 33-Nation Study. *Science* **332**:1100–1104. [12]

Ghate, R., and H. Nagendra. 2005. Role of Monitoring in Institutional Performance: Forest Management in Maharashtra, India. *Conserv. Soc.* **3**:509–532. [5]

Gintis, H. 2010. Social Norms as Choreography. *Polit. Philos. Econ.* **9**:251–264. [12]

Giraldeau, L.-A., and G. Beauchamp. 1999. Food Exploitation: Searching for the Optimal Joining Policy. *Trends Ecol. Evol.* **14**:102–106. [12]

Giraldeau, L.-A., and T. Caraco. 2000. Social Foraging Theory. Monographs in Behavior and Ecology. Princeton: Princeton Univ. Press. [3, 4]

Giraldeau, L.-A., and F. Dubois. 2008. Social Foraging and the Study of Exploitative Behavior. *Adv. Study Behav.* **38**:59–104. [3, 4, 11]

Giraldeau, L.-A., T. J. Valone, and J. J. Templeton. 2002. Potential Disadvantages of Using Socially Acquired Information. *Phil. Trans. R. Soc. B* **357**:1559–1566. [11]

Giraud, T., J. S. Pedersen, and L. Keller. 2002. Evolution of Supercolonies: The Argentine Ants of Southern Europe. *PNAS* **99**:6075–6079. [10]

Glaeser, E., B. Sacerdote, and J. Scheinkman. 1996. Crime and Social Interactions. *Q. J. Econ.* **111**:507–548. [2]

Goette, L., D. Huffman, and S. Meier. 2006. The Impact of Group Membership on Cooperation and Norm Enforcement: Evidence Using Random Assignment to Social Groups. *Am. Econ. Rev.* **96**:212–216. [10]

Goldberg, J. L., J. W. A. Grant, and L. Lefebvre. 2001. Effects of the Temporal Predictability and Spatial Clumpig of Food on the Intensity of Competitive Aggression in Zenadia Dove. *Behav. Ecol.* **12**:490–495. [11]

Goodale, E., G. Beauchamp, R. D. Magrath, J. C. Nieh, and G. D. Ruxton. 2010. Interspecific Information Transfer Influences Animal Community Structure. *Trends Ecol. Evol.* **25**:354–361. [11]

Gough, E., H. Shaikh, and A. R. Manges. 2011. Systematic Review of Intestinal Microbiota Transplantation (Fecal Bacteriotherapy) for Recurrent *Clostridium difficile* Infection. *Clin. Infect. Dis.* **53**:994–1002. [7]

Goyal, S. 2007. Connections: An Introduction to the Economics of Networks. Princeton: Princeton Univ. Press. [2]

Grafen, A. 2006. Optimization of Inclusion Fitness. *J. Theor. Biol.* **238**:541–563. [3]

Grechenig, K., A. Nicklisch , and C. Thöni. 2010. Punishment Despite Reasonable Doubt: A Public Goods Experiment with Sanctions under Uncertainty. *J. Empir. Leg. Stud.* **7**:847–867. [10]

Green, E. J., and R. H. Porter. 1984. Noncooperative Collusion under Imperfect Price Information. *Econometrica* **52**:87–100. [2]

Griffin, A. S., S. A. West, and A. Buckling. 2004. Cooperation and Competition in Pathogenic Bacteria. *Nature* **430**:1024–1027. [7]

Grimm, V., and F. Mengel. 2012. An Experiment on Learning in a Multiple Games Environment. *J. Econ. Theory* **147**:2220–2259. [2]

Gross, M. R. 1996. Alternative Reproductive Strategies and Tactics: Diversity within Sexes. *Trends Ecol. Evol.* **11**:92–98. [12]

Grzelak, J. L., M. Poppe, Z. Czwartosz, and A. Nowak. 1988. Numerical Trap: A New Look at Outcome Representation in Studies on Choice Behavior. *Eur. J. Soc. Psychol.* **18**:143–159. [3]

Guala, F., L. Mittone, and M. Ploner. 2013. Group Membership, Team Preferences, and Expectations. *J. Econ. Behav. Organ.* **86**:183–190. [12]

Guererk, O., B. Irlenbusch, and B. Rockenbach. 2006. The Competitive Advantage of Sanctioning Institutions. *Science* **312**:108-111. [2]

Gulati, R. 1998. Alliances and Networks. *Strategic Manag. J.* **19**:293–317. [12]

Guth, W., and M. G. Kocher. 2014. More Than Thirty Years of Ultimatum Bargaining Experiments: Motives, Variations, and a Survey of the Recent Literature. *J. Econ. Behav. Organ.* **108**:396–409. [3]

Hajjar, R., J. A. Oldekop, A. Agrawal, et al. 2016. The Data Not Collected on Community Forestry. *Conserv. Biol.* DOI: 10.1111/cobi.12732. [5]

Hamblin, S., and L.-A. Giraldeau. 2009. Finding the Evolutionarily Stable Learning Rule for Frequency-Dependent Foraging. *Anim. Behav.* **78**:1343–1350. [3]

Hamblin, S., K. J. Mathot, J. Morand-Ferron, et al. 2010. Predator Inadvertent Social Information Use Favours Reduced Clumping of Its Prey. *Oikos* **119**:286–291. [11]

Hamilton, I. M. 2002. Kleptoparasitism and the Distribution of Unequal Competitors. *Behav. Ecol.* **13**:260–267. [3, 4]

Hamilton, W. D. 1964. The Genetical Evolution of Social Behaviour: I and II. *J. Theor. Biol.* **7**:1–52. [3, 4, 7]

Hansen, M. C., P. V. Potapov, R. Moore, et al. 2013. High-Resolution Global Maps of 21st-Century Forest Cover Change. *Science* **342**:850–853. [5]

Hardin, G. 1968. The Tragedy of the Commons. *Science* **162**:1243. [5, 6, 8, 9, 12]

Hare, B., J. Call, B. Agnetta, and M. Tomasello. 2000. Chimpanzees Know What Conspecifics Do and Do Not See. *Anim. Behav.* **59**:771–785. [11]

Harley, C. B. 1981. Learning the Evolutionary Stable Strategies. *J. Theor. Biol.* **89**:611–633. [3, 11]

Hart, K. 1975. Swindler or Public Benefactor? The Entrepreneur in His Community. In: Changing Social Structure in Ghana, ed. J. Good, pp. 1–35. London: International African Institute. [3]

Hartner, M., S. Rechberger, E. Kirchler, and A. Schabmann. 2008. Procedural Fairness and Tax Compliance. *Econ. Anal. Policy* **38**:137–152. [12]

Hede, K. 2014. Antibiotic Resistance: An Infectious Arms Race. *Nature* **509**:S2–3. [9]

Heinrich, B., and J. M. Marzluff. 1991. Do Common Ravens Yell Because They Want to Attract Others? *Behav. Ecol. Sociobiol.* **28**:13–21. [11]

Heller, Y. 2015. Three Steps Ahead. *Theor. Econ.* **10**:203–241. [3]

Henrich, B., and J. W. Pepper. 1998. Influence of Competitors on Caching Behaviour in the Common Raven, *Corvus corax. Anim. Behav.* **56**:1083–1090. [11]

Henrich, J. 2004. Cultural Group Selection, Coevolutionary Processes and Large-Scale Cooperation. *J. Econ. Behav. Organ.* **53**:3–35. [3]

Henrich, J., J. Ensminger, R. McElreath, et al. 2010. Markets, Religion, Community Size, and the Evolution of Fairness and Punishment. *Science* **327**:1480–1484. [12]

Hentzer, M., H. Wu, J. B. Andersen, et al. 2003. Attenuation of *Pseudomonas aeruginosa* Virulence by Quorum Sensing Inhibitors. *EMBO J.* **22**:3803–3815. [7]

Herrmann, B., C. Thoeni, and S. Gaechter. 2008. Antisocial Punishment across Societies. *Science* **319**:1362–1367. [2]

Herrmann, M. 2010. Monopoly Pricing of an Antibiotic Subject to Bacterial Resistance. *J. Health Econ.* **29**:137–150. [8, 9]

Herrmann, M., and G. Gaudet. 2009. The Economic Dynamics of Antibiotic Efficacy under Open Access. *J. Environ. Econ. Manage.* **57**:334–350. [8, 9]

Herrmann, M., and R. Laxminarayan. 2010. Antibiotic Effectiveness: New Challenges in Natural Resource Management. *Annu. Rev. Res. Econ.* **2**:125–138. [8]

Herrmann, M., B. Nkuiya, and A.-R. Dussault. 2013. Innovation and Antibiotic Use within Antibiotic Classes: Market Incentives and Economic Instruments. *Resource Ener. Econ.* **50**:267–284. [8]

Hijmans, R. J., S. E. Cameron, J. L. Parra, P. G. Jones, and A. Jarvis. 2005. Very High Resolution Interpolated Climate Surfaces for Global Land Areas. *Int. J. Climatol.* **25**:1965–1978. [5]

Hilbe, C., and K. Sigmund. 2010. Incentives and Opportunism: From the Carrot to the Stick. *Proc. R. Soc. Lond. B* **277**:2427–2433. [10]

Hilborn, R., and C. J. Walters, eds. 1992. Quantitative Fisheries Stock Assessment: Choice, Dynamics and Uncertainty. London: Chapman & Hall. [6]

Hirshleifer, J. 1989. Conflict and Rent-Seeking Success Functions: Ratio Vs Difference Models of Relative Success. *Public Choice* **63**:101–112. [12]

———. 1991. The Technology of Conáict as an Economic Activity. *Am. Econ. Rev.* **81**:130–134. [12]

Hoare, D. J., J. Krause, N. Peuhkuri, and J. G. J. Godin. 2000a. Body Size and Shoaling in Fish. *J. Fish Biol.* **57**:1351–1366. [12]

Hoare, D. J., G. D. Ruxton, J. G. J. Godin, and J. Krause. 2000b. The Social Organization of Free-Ranging Fish Shoals. *Oikos* **89**:546–554. [12]

Hopkins, E. 2002. Two Competing Models of How People Learn in Games. *Econometrica* **70**:2141–2166. [3]

Hoppitt, W., and K. N. Laland. 2011. Detecting Social Learning Using Networks: A Users Guide. *Am. J. Primatol.* **73**:834–844. [12]

Houston, A. I., and J. M. M. McNamara. 1999. Models of Adaptive Behaviour. Cambridge: Cambridge Univ. Press. [4]

Howard, D. H. 2004. Resistance-Induced Antibiotic Substitution. *Health Econ.* **13**:585–595. [8]

———. 2005. Life Expectancy and the Value of Early Detection. *J. Health Econ.* **24**:891–906. [8]

Huck, S. 1998. Trust, Treason, and Trials: An Example of How the Evolution of Preferences Can Be Driven by Legal Institutions. *J. Law Econ. Organ.* **14**:44–60. [2]

Huck, S., and M. Kosfeld. 2007. The Dynamics of Neighbourhood Watch and Norm Enforcement. *Econ. J.* **117**:270–286. [2]

Hughes, W. O. H., and J. J. Boomsma. 2008. Genetic Royal Cheats in Leaf-Cutting Ant Societies. *PNAS* **105**:5150–5153. [10]

Hurwicz, L., and S. Reiter. 2006. Designing Economic Mechanisms. Cambridge: Cambridge Univ. Press. [3]

Hwang, S.-H. 2009. Contest Success Functions: Theory and Evidence. Economics Department Working Paper Series, Paper Nr. 11. Amherst: Univ. of Massachusetts-Amherst. [12]

Isaac, R. M., J. M. Walker, and S. H. Thomas. 1984. Divergent Evidence on Free Riding: An Experimental Examination of Possible Explanations. *Public Choice* **43**:113–149. [3]

Isbell, L. A. 1991. Contest and Scramble Competition: Patterns of Female Aggression and Ranging Behavior among Primates. *Behav. Ecol.* **2**:143–155. [12]

Jacquet, J., and D. Pauly. 2008. Funding Priorities: Big Barriers to Small-Scale Fisheries. *Conserv. Biol.* **22**:832–835. [6]

Janssen, M. A., R. Holahan, A. Lee, and E. Ostrom. 2010. Lab Experiments for the Study of Social-Ecological Systems. *Science* **328**:613–617. [4]

Jehiel, P. 2001. Limited Foresight May Force Cooperation. *Rev. Econ. Stud.* **68**:369–391. [3]

———. 2005. Analogy-Based Expectation Equilibrium. *J. Econ. Theory* **123**:81–104. [2, 3]

Jenkins, R. K. B., A. Keane, A. R. Rakotoarivelo, et al. 2011. Analysis Patterns of Bushmeat Consumption Reveals Extensive Exploitation of Protected Species in Eastern Madagascar. *PLoS One* **6**:e27570. [5]

Jenkins, S. H., A. Rothstein, and W. C. H. Green. 1995. Food Hoarding by Merriam's Kangaroo Rats: A Test of Alternative Hypotheses. *Ecology* **76**:2470–2481. [11]

Johnson, C. 2001. Community Formation and Fisheries Conservation in Southern Thailand. *Dev. Change* **32**:951–974. [5]

Johnstone, R. A. 2001. Eavesdropping and Animal Conflict. *PNAS* **98**:9177–9180. [11]

Johnstone, R. A., and L. A. Dugatkin. 2000. Coalition Formation in Animals and the Nature of Winner and Loser Effects. *Proc. R. Soc. Lond. B* **267**:17–21. [10]

Jones, G. P. 1983. Relationship between Density and Behaviour in Juvenile *Pseudolabrus celidotus* (Pisce: Labridae). *Anim. Behav.* **31**:729–735. [11]

Kacelnik, A. 2006. Meanings of Rationality. In: Rational Animals?, ed. S. N. Hurley, M., pp. 87–106. New York: Oxford Univ. Press. [3]

Kameda, T., and D. Nakanishi. 2002. Cost-Benefit Analysis of Social/Cultural Learning in a Nonstationary Uncertain Environment: An Evolutionary Simulation and an Experiment with Human Subjects. *Evol. Hum. Behav.* **23**:373–393. [12]

Kameda, T., T. Tsukasaki, R. Hastie, and N. Berg. 2011. Democracy under Uncertainty: The Wisdom of Crowds and the Free-Rider Problem in Group Decision Making. *Psychol. Rev.* **118**:76–96. [12]

Karandikar, R., D. Mokherjee, D. Ray, and F. V. Redondo. 1998. Evolving Aspirations and Cooperation. *J. Econ. Theory* **80**:292–331. [2]

Katsnelson, E., U. Motro, M. W. Feldman, and A. Lotem. 2008. Early Experience Affects Producer-Scrounger Foraging Tendencies in the House Sparrow. *Anim. Behav.* **75**:1465–1472. [3]

———. 2011. Individual-Learning Ability Predicts Social-Foraging Strategy in House Sparrows. *Proc. R. Soc. Lond. B* **278**:582–589. [3]

Keeling, M. J., and K. T. D. Eames. 2005. Networks and Epidemic Models. *J. R. Soc. Interface* **2**:295–307. [12]

Kendal, J., L.-A. Giraldeau, and K. Laland. 2009. The Evolution of Social Learning Rules: Payoff-Biased and Frequency-Dependent Biased Transmission. *J. Theor. Biol.* **260**:210–219. [3]

Kerth, G. 2010. Group Decision-Making in Animal Societies. In: Animal Behaviour: Evolution and Mechanisms, ed. P. Kappeler, pp. 241–265. Berlin: Springer [12]

Kerth, G., C. Ebert, and C. Schmidtke. 2006. Group Decision Making in Fission-Fusion Societies: Evidence from Two-Field Experiments in Bechstein's Bats. *Proc. R. Soc. Lond. B* **273**:2785–2790. [12]

Kiester, A. R., T. Nagylaki, and B. Shaffer. 1981. Population Dynamics of Species with Synogenetic Sibling Species. *Theor. Popul. Biol.* **19**:358–369. [6]

Kilner, R. M., and N. E. Langmore. 2011. Cuckoos versus Hosts in Insects and Birds: Adaptations, Counter-Adaptations and Outcomes. *Biological Reviews* **86**:836–852. [12]

King, A. J., F. E. Clark, and G. Cowlishaw. 2011. The Dining Etiquette of Desert Baboons: The Roles of Social Bonds, Kinship, and Dominance in Co-Feeding Networks. *Am. J. Primatol.* **73**:768–774. [12]

King, A. J., and G. Cowlishaw. 2007. When to Use Social Information: The Advantage of Large Group Size in Individual Decision Making. *Biol. Lett.* **3**:137–139. [12]

King, A. J., C. M. S. Douglas, E. Huchard, N. J. B. Isaac, and G. Cowlishaw. 2008. Dominance and Affiliation Mediate Despotism in a Social Primate. *Curr. Biol.* **18**:1833–1838. [12]

King, A. J., N. J. B. Isaac, and G. Cowlishaw. 2009. Ecological, Social, and Reproductive Factors Shape Producer-Scrounger Dynamics in Baboons. *Behav. Ecol.* **20**:1039–1049. [12]

King, A. J., and C. Sueur. 2011. Where Next? Group Coordination and Collective Decision Making by Primates. *Int. J. Primatol.* **32**:1245–1267. [12]

King, A. J., L. J. Williams, and C. Mettke-Hofmann. 2015. The Effects of Social Conformity on Gouldian Finch Personality. *Anim. Behav.* **99**:25–31. [12]

Koita, O. A., O. K. Doumbo, A. Ouattara, et al. 2012. False-Negative Rapid Diagnostic Tests for Malaria and Deletion of the Histidine-Rich Repeat Region of the Hrp2 Gene. *Am. J. Trop. Med. Hyg.* **86**:194–198. [9]

Kokko, H., K. U. Heubel, and D. J. Rankin. 2008. How Populations Persist When Asexuality Requires Sex: The Spatial Dynamics of Coping with Sperm Parasites. *Proc. R. Soc. Lond. B* **275**: 817–825. [6]

Kokko, H., R. A. Johnstone, and T. H. Clutton-Brock. 2001. The Evolution of Cooperative Breeding through Group Augmentation. *Proc. R. Soc. Lond. B* **268**:187–196. [3]

Koops, M. A., and L.-A. Giraldeau. 1996. Producer-Scrounger Foraging Games in Starlings: A Test of Rate-Maximizing and Risk-Sensitive Models. *Anim. Behav.* **51**:773–783. [3]

Kosfeld, M., and S. Neckermann. 2011. Getting More Work for Nothing? Symbolic Awards and Worker Performance. *Am. Econ. J. Microecon.* **3**:86–99. [3]

Kosfeld, M., A. Okada, and A. Riedl. 2009. Institution Formation in Public Goods Games. *Am. Econ. Rev.* **99**:1335–1355. [2, 12]

Kosfeld, M., and D. Rustagi. 2015. Leader Punishment and Cooperation in Groups: Experimental Field Evidence from Commons Management in Ethiopia. *Am. Econ. Rev.* **105**:747–783. [5, 6]

Kreps, D. M. 1988. Notes on the Theory of Choice. Boulder: Westview Press. [3]

Kreps, D. M., P. Milgrom, J. Roberts, and R. Wilson. 1982. Rational Cooperation in the Finitely Dilemma Repeated Prisoners' Dilemma. *J. Econ. Theory* **252**:245–252. [2, 3]

Krueger, A. B., and A. Mas. 2004. Strikes, Scabs, and Tread Separations: Labor Strife and the Production of Defective Bridgestone/Firestone Tires. *J. Polit. Econ.* **112**:253–289. [10]

Kruuk, H. 1972. The Spotted Hyena. Chicago: Univ. of Chicago Press. [11]

Kuhn, H. W., J. C. Harsanyi, R. Selten, et al. 1996. The Work of John Nash in Game Theory: Nobel Seminar, Dec. 8, 1994. *J. Econ. Theory* **69**:153–185. [3]

Kummerli, R., and S. P. Brown. 2010. Molecular and Regulatory Properties of a Public Good Shape the Evolution of Cooperation. *PNAS* **107**:18921–18926. [7]

Kummerli, R., and A. Ross-Gillespie. 2014. Explaining the Sociobiology of Pyoverdin Producing *Pseudomonas*: A Comment on Zhang and Rainey (2013). *Evolution* **68**:3337–3343. [7]

Kupper, C., M. Stocks, J. E. Risse, et al. 2016. A Supergene Determines Highly Divergent Male Reproductive Morphs in the Ruff. *Nat. Genet.* **48**:79–83. [3]

Lahti, K., K. Koivula, S. Rytkonen, et al. 1998. Social Influences on Food Caching in Willow Tits: A Field Experiment. *Behav. Ecol.* **9**:122–129. [11]

Lange, A., and R. Dukas. 2009. Bayesian Approximations and Extensions: Optimal Decisions for Small Brains and Possibly Big Ones Too. *J. Theor. Biol.* **259**:503–516. [3]

Lank, D. B., C. M. Smith, O. Hanotte, T. Burke, and F. Cooke. 1995. Genetic Polymorphism for Alternative Mating-Behavior in Lekking Male Ruff *Philomachus pugnax*. *Nature* **378**:59–62. [3]

Laskowski, K. L., and J. N. Pruitt. 2014. Evidence of Social Niche Construction: Persistent and Repeated Social Interactions Generate Stronger Personalities in a Social Spider. *Proc. R. Soc. Lond. B* **281**:7. [12]

Laxminarayan, R., and G. M. Brown. 2001. Economics of Antibiotic Resistance: A Theory of Optimal Use. *J. Environ. Econ. Manage.* **42**:183–206. [8]

Laxminarayan, R., and M. Herrmann. 2015. Biological Resistance. In: Handbook on the Economics of Natural Resources, ed. R. Halvorsen and D. F. Layton, pp. 249–278. Northampton, MA: Edward Elgar. [8]

Laxminarayan, R., A. Malani, D. H. Howard, and D. L. Smith. 2007. Extending the Cure: Policy Responses to the Growing Threat of Antibiotic Resistance. Washington, D.C.: Resources for the Future. [8]

Laxminarayan, R., and D. L. Smith. 2006. Ecology and Econimics of Cycling Antibiotics: Insights from Mathematical Modeling. In: Multiple Drug Resistant Bacteria, ed. C.-A. Cuevas, pp. 167–178. Norwich: Horizon Scientific Press. [8]

Laxminarayan, R., and M. L. Weitzman. 2002. On the Implications of Endogenous Resistance to Medications. *J. Health Econ.* **21**:709–718. [8]

Leaver, L. A., L. Hopewell, C. Caldwell, and L. Mallarky. 2007. Audience Effects on Food Caching in Grey Squirrels (*Sciurus carolinensis*): Evidence for Pilferage Avoidance Strategies. *Anim. Cogn.* **10**:23–27. [11]

Ledyard, J. O. 1995. Public Goods: A Survey of Experimental Research. In: Handbook of Experimental Economics, ed. J. H. Kagel and A. E. Roth, pp. 111–194. Princeton: Princeton Univ. Press. [2, 6,10]

Leggett, H. C., S. P. Brown, and S. E. Reece. 2014. War and Peace: Social Interactions in Infections. *Phil. Trans. R. Soc. B* **369**: [7, 9]

Lehmann, L., and M. W. Feldman. 2009. Coevolution of Adaptive Technology, Maladaptive Culture and Population Size in a Producer-Scrounger Game. *Proc. R. Soc. Lond. B* **276**:3853–3862. [3]

Lehmann, L., L. Keller, S. West, and D. Roze. 2007. Group Selection and Kin Selection: Two Concepts but One Process. *PNAS* **104**:6736–6739. [3]

Le Kama, A. A., and K. Schubert. 2007. A Note on the Consequences of an Endogenous Discounting Depending on the Environmental Quality. *Macroecon. Dyn.* **11**:272–289. [9]

Lendvai, A. Z., Z. Barta, A. Liker, and V. Bokony. 2004. The Effect of Energy Reserves on Social Foraging: Hungry Sparrows Scrounge More. *Proc. R. Soc. Lond. B* **271**:2467–2472. [3]

Levin, B. R., and J. J. Bull. 2004. Population and Evolutionary Dynamics of Phage Therapy. *Nat. Rev. Microbiol.* **2**:166–173. [7]

Lewis, S. L., and M. A. Maslin. 2015. Defining the Anthropocene. *Nature* **519**:171–180. [4]

Liebrand, W. B. G., and C. G. McClintock. 1988. The Ring Measure of Social Values - Computerized Procedure for Assessing Individual-Differences in Information-Processing and Social Value Orientation. *Eur. J. Personality* **2**:217–230. [3]

Lindbeck, A., S. Nyberg, and J. Weibull. 1999. Social Norms and Economic Incentives in the Welfare State. *Q. J. Econ.* **CXIV**:1–35. [2]

Lotem, A., M. A. Fishman, and L. Stone. 1999. Evolution of Cooperation between Individuals. *Nature* **400**:226–227. [2]

Loukola, O. J., T. Laaksonen, J.-T. Seppänen, and J. T. Forsman. 2014. Active Hiding of Social Information from Information-Parasites. *BMC Evol. Biol.* **14**:32. [11]

Lucas, J. R., R. D. Howard, and J. G. Palmer. 1996. Callers and Satellites: Chorus Behaviour in Anurans as a Stochastic Dynamic Game. *Anim. Behav.* **51**:501–518. [3]

Maeda, T., R. Garcia-Contreras, M. Pu, et al. 2012. Quorum Quenching Quandary: Resistance to Antivirulence Compounds. *ISME J.* **6**:493–501. [7]

Manson, J. H., and R. W. Wrangham. 1991. Intergroup Aggression in Chimpanzees and Humans. *Curr. Anthropol.* **32**:369–390. [12]

Marshall, J. A. R. 2011. Group Selection and Kin Selection: Formally Equivalent Approaches. *Trends Ecol. Evol.* **26**:325–332. [3]

Mas, A. 2008. Labour Unrest and the Quality of Production: Evidence from the Construction Equipment Resale Market. *Rev. Econ. Stud.* **75**:229–258. [10]

Masiero, G., M. Filippini, M. Ferech, and H. Goossens. 2010. Socioeconomic Determinants of Outpatient Antibiotic Use in Euope. *Int. J. Pub. Health* **55**:469–478. [8]

Mathot, K. J., and L.-A. Giraldeau. 2010. Within-Group Relatedness Can Lead to Higher Levels of Exploitation: A Model and Empirical Test. *Behav. Ecol.* **21**:843–850. [3, 4, 10]

Mathot, K. J., S. Godde, V. Careau, D. W. Thomas, and L.-A. Giraldeau. 2009. Testing Dynamic Variance-Sensitive Foraging Using Individual Differences in Basal Metabolic Rates of Zebra Finches. *Oikos* **118**:545–552. [3]

Matsumura, K., S. Matsunaga, and N. Fusetani. 2007. Phosphatidylcholine Profile-Mediated Group Recognition in Catfish. *J. Exp. Biol.* **210**:1992–1999. [12]

Maynard-Smith, J. 1982. Evolution and the Theory of Games. Cambridge: Cambridge Univ. Press. [3, 11]

Maynard-Smith, J., and G. R. Price. 1973. The Logic of Animal Conflict. *Nature* **246**:15–18. [2, 3, 11]

McCormack, J. E., P. G. Jablonski, and J. L. Brown. 2007. Producer-Scrounger Roles and Joining Based on Dominance in a Free-Living Group of Mexican Jays (*Aphelocoma ultramarina*). *Behaviour* **144**:967–982. [3]

McGinty, S. E., D. J. Rankin, and S. P. Brown. 2011. Horizontal Gene Transfer and the Evolution of Bacterial Cooperation. *Evolution* **65**:21–32. [10]

McNally, L., and S. P. Brown. 2015. Building the Microbiome in Health and Disease: Niche Construction and Social Conflict in Bacteria. *Phil. Trans. R. Soc. B* **370**: [7]

McNally, L., M. Viana, and S. P. Brown. 2014. Cooperative Secretions Facilitate Host Range Expansion in Bacteria. *Nat. Commun.* **5**:4594. [7]

McNamara, J. M. 2013. Towards a Richer Evolutionary Game Theory. *J. R. Soc. Interface* **10**:20130544. [3, 11]

McNamara, J. M. and A. I. Houston. 1989. State-Dependent Contests for Food. *J. Theor. Biol.* **137**:457–479. [4]

McNamara, J. M., P. A. Stephens, S. R. X. Dall, and A. I. Houston. 2009. Evolution of Trust and Trustworthiness: Social Awareness Favours Personality Differences. *Proc. Biol. Sci.* **276**:605–613. [11]

Mechoulan, S. 2007. Market Structure and Communicable Diseases. *Can. J. Econ.* **40**:468–492. [8]

Mellbye, B., and M. Schuster. 2011. The Sociomicrobiology of Antivirulence Drug Resistance: A Proof of Concept. *Mbio* **2**: [7]

Mellström, C., and M. Johannesson. 2008. Crowding out in Blood Donation: Was Titmuss Right? *J. Eur. Econ. Assoc.* **6**:845–863. [10]

Mengel, F. 2008. Matching Structure and the Cultural Transmission of Social Norms. *J. Econ. Behav. Organ.* **67**:608–623. [2, 12]

———. 2009. Conformism and Cooperation in a Local Interaction Model. *J. Evol. Econ.* **19**:397–415. [2]

———. 2012. Learning across Games. *Games Econ. Behav.* **74**:601–619. [2, 3]

Mesoudi, A. 2011. Cultural Evolution How Darwinian Theory Can Explain Human Culture and Synthesize the Social Sciences. Chicago: Univ. of Chicago Press. [3]

Mesterton-Gibbons, M., S. Gavrilets, J. Gravner, and E. Akcay. 2011. Models of Coalition or Alliance Formation. *J. Theor. Biol.* **274**:187–204. [10]

Meyfroidt, P., and E. F. Lambin. 2011. Global Forest Transition: Prospects for an End to Deforestation. *Annu. Rev. Environ. Resour.* **36**:343–371. [5]

Milinski, M., C. Hilbe, D. Semmann, R. Sommerfeld, and J. Marotzke. 2016. Humans Choose Representatives Who Enforce Cooperation in Social Dilemmas through Extortion. *Nat. Commun.* **7**:10915 [3]

Molnar, A., D. Gomes, R. Sousa, et al. 2008. Community Forest Enterprise Markets in Mexico and Brazil: New Opportunities and Challenges for Legal Access to the Forest. *J. Sust. Forestry* **27**:87–121. [5]

Moore, J. 2002. Parasites and the Behavior of Animals. Oxford Series in Ecology and Evolution. New York: Oxford Univ. Press. [3]

Morand-Ferron, J., and L.-A. Giraldeau. 2010. Learning Behaviorally Stable Solutions to Producer-Scrounger Games. *Behav. Ecol.* **21**:343–348. [3, 12]

Morand-Ferron, J., L.-A. Giraldeau, and L. Lefebvre. 2007. Wild Carib Grackles Play a Producer-Scrounger Game. *Behav. Ecol.* **18**:916–921. [3]

Morand-Ferron, J., G. M. Wu, and L.-A. Giraldeau. 2011. Persistent Individual Differences in Tactic Use in a Producer-Scrounger Game Are Group Dependent. *Anim. Behav.* **82**:811–816. [3]

Morgan, A. D., B. J. Z. Quigley, S. P. Brown, and A. Buckling. 2012. Selection on Non-Social Traits Limits the Invasion of Social Cheats. *Ecol. Lett.* **15**:841–846. [7]

Mossong, J., N. Hens, M. Jit, et al. 2008. Social Contacts and Mixing Patterns Relevant to the Spread of Infectious Diseases. *PloS Medicine* **5**:381–391. [12]

Mottley, K., and L.-A. Giraldeau. 2000. Experimental Evidence That Group Foragers Can Converge on Predicted Producer-Scrounger Equilibria. *Anim. Behav.* **60**:341–350. [3]

Munro, G. R. 1979. The Optimal Management of Transboundary Renewable Resources. *Can. J. Econ.* **12**:355–376. [6]

Murphy, K. M., A. Shleifer, and R. Vishny. 1991. The Allocation of Talent: Implications for Growth. *Q. J. Econ.* **106**:503–530. [10]

———. 1993. Why Is Rent-Seeking So Costly to Growth. *Am. Econ. Rev.* **83**:409–414. [3]

Murphy, R. O., and K. A. Ackermann. 2014. Social Value Orientation: Theoretical and Measurement Issues in the Study of Social Preferences. *Pers. Soc. Psychol. Rev.* **18**:13–41. [3]

Murphy, R. O., K. A. Ackermann, and M. J. J. Handgraaf. 2011. Measuring Social Value Orientation. *Judgm. Decis. Mak.* **6**:771–781. [3]

Murray, C. K., P. Frijters, and M. Vorster. 2015. Give and You Shall Receive: Emergence of Loyalty and Groups in a Repeated Rent-Allocation Game. Secondary Give and You Shall Receive: Emergence of Loyalty and Groups in a Repeated Rent-Allocation Game. http://ftp.iza.org/dp9010.pdf. (accessed Aug. 25, 2016). [10]

Nafziger, E. W. 1969. Effect of Nigerian Extended Family on Entrepreneurial Activity. *Econ. Dev. Cult. Change* **18**:25–33. [3]

Nagendra, H., M. Karmacharya, and B. Karna. 2005. Evaluating Forest Management in Nepal: Views across Space and Time. *Ecol. Soc.* **10**:24–40. [5]

Nash, J. F. 1950. Equilibrium Points in N-Person Games. *PNAS* **36**:48–49. [3]

Nax, H. H., M. N. Burton-Chellew, S. A. West, and H. P. Young. 2016. Learning in a Black Box. *J. Econ. Behav. Organ.* **127**:1–15. [3]

Newman, M. E. J. 2002. Spread of Epidemic Disease on Networks. *Physical Review E* **66**:016128 [12]

Newton, P., J. A. Oldekop, G. Brodnig, B. Karna, and A. Agrawal. 2016. Carbon, Biodiversity, and Livelihoods in Forest Commons: Synergies, Trade-Offs, and Implications for REDD+. *Environmental Research Letters* **11**:044017. [5]

Nogueira, T., D. J. Rankin, M. Touchon, et al. 2009. Horizontal Gene Transfer of the Secretome Drives the Evolution of Bacterial Cooperation and Virulence. *Curr. Biol.* **19**:1683–1691. [7]

Noldeke, G., and L. Samuelson. 1993. An Evolutionary Analysis of Backward and Forward Induction. *Games Econ. Behav.* **5**:425–454. [3]

Nowak, M. A., and K. Sigmund. 1998. Evolution of Indirect Reciprocity by Image Scoring. *Nature* **393**:573–577. [2]

Nowak, M. A., C. E. Tarnita, and E. O. Wilson. 2010. The Evolution of Eusociality. *Nature* **466**:1057–1062. [3]

Ochman, H., J. G. Lawrence, and E. A. Groisman. 2000. Lateral Gene Transfer and the Nature of Bacterial Innovation. *Nature* **405**:299–304. [12]

Ockenfels, A., and P. Werner. 2014. Beliefs and Ingroup Favoritism. *J. Econ. Behav. Organ.* **108**:453–462. [12]

Ohtsuki, H., C. Hauert, E. Lieberman, and M. A. Nowak. 2006. A Simple Rule for the Evolution of Cooperation on Graphs and Social Networks. *Nature* **441**:502–505. [12]

Oldekop, J. A., A. J. Bebbington, D. Brockington, and R. F. Preziosi. 2010. Understanding the Lessons and Limitations of Conservation and Development. *Conserv. Biol. Pract.* **24**:461–469. [5]

Oldekop, J. A., A. J. Bebbington, K. Hennermann, et al. 2013. Evaluating the Effects of Common Pool Resource Institutions and Market Forces on Species Richness and Forest Cover in Ecuadorian Indigenous Kichwa Communities. *Conserv. Lett.* **6**:107–115. [5]

Oldekop, J. A., A. J. Bebbington, N. K. Truelove, et al. 2012. Environmental Impacts and Scarcity Perception Influence Local Institutions in Indigenous Amazonian Kichwa Communities. *Hum. Ecol.* **40**:101–115. [5]

Oldekop, J. A., L. B. Fontana, J. Grugel, et al. 2016. 100 Key Research Questions for the Post-2015 Development Agenda. *Dev. Policy Rev.* **34**:55–82. [5]

Oliver, P. E. 1993. Formal Models of Collective Action. *Annu. Rev. Sociol.* **19**:271–300. [12]

Olson, M. 1965. The Logic of Collective Action : Public Goods and the Theory of Groups. Harvard Economic Studies. Cambridge, MA: Harvard Univ. Press. [3]

Ostrom, E. 1990. Governing the Commons: The Evolution of Institutions for Collective Action. Cambridge: Cambridge Univ. Press. [4–6]

———. 2007. A Diagnostic Approach for Going Beyond Panaceas. *PNAS* **104**:15181–15187. [3]

———. 2009. A General Framework for Analyzing Sustainability of Social-Ecological Systems. *Science* **325**:419–422. [5]

Ostrom, E., and H. Nagendra. 2006. Insights on Linking Forests, Trees, and People from the Air, on the Ground, and in the Laboratory *PNAS* **103**:19224–19231. [6]

Ostrom, E., J. Walker, and R. Gardner. 1992. Covenants with and without a Sword: Self-Governance Is Possible. *Am. Polit. Sci. Rev.* **86**:404–417. [10]

Oyono, P. R. 2005. Profiling Local-Level Outcomes of Environmental Decentralizations: The Case of Cameroon's Forests in the Congo Basin. *J. Env. Develop.* **14**:317–337. [5]

Pagdee, A., Y. Kim, and P. J. Daugherty. 2006. What Makes Community Forest Management Successful: A Meta-Study from Community Forests Throughout the World. *Soc. Nat. Resour.* **19**:33–52. [5]

Pailler, S., R. Naidoo, N. D. Burgess, O. E. Freeman, and B. Fisher. 2015. Impacts of Community-Based Natural Resource Management on Wealth, Food Security and Child Health in Tanzania. *PLoS One* **10**:e0133252. [5]

Pammolli, L., L. Magazzini, and M. Riccaboni. 2011. The Productivity Crisis in Pharmaceutical R&D. *Nature Rev. Drug Disc.* **10**:428–438. [8]

Pauly, D., V. Christensen, S. Guénette, et al. 2002. Towards Sustainability in World Fisheries. *Nature* **418**:689–695. [6]

Perc, M., J. Gomez-Gardenes, A. Szolnoki, L. M. Floria, and Y. Moreno. 2013. Evolutionary Dynamics of Group Interactions on Structured Populations: A Review. *J. R. Soc. Interface* **10**:20120997. [12]

Persha, L., A. Agrawal, and A. Chhatre. 2011. Social and Ecological Synergy: Local Rulemaking, Forest Livelihoods and Biodiversity Conservation. *Science* **331**:1606–1608. [5]

Pettit, P., and R. Sugden. 1989. The Backward Induction Paradox. *J. Philos.* **86**:169–182. [3]

Pollnac, R. B., B. R. Crawford, and M. L. G. Gorospe. 2001. Discovering Factors That Influence the Success of Community-Based Marine Protected Areas in the Visayas, Philippines. *Ocean Coast. Manage.* **44**:683–710. [6]

Pollock, G. B., Cabrales, A., and S. W. Rissing. 2004. On Suicidal Punishment among *Acromyrmex versicolor* Co-Foundresses: The Disadvantage in Personal Advantage. *Evol. Ecol. Res.* **6**:891–917. [6]

Pollock, M. R. 1967. Origin and Function of Penicillinase: A Problem in Biochemical Evolution. *Br. Med. J.* **4**:71–77. [9]

Popat, R., D. M. Cornforth, L. McNally, and S. P. Brown. 2015. Collective Sensing and Collective Responses in Quorum-Sensing Bacteria. *J. R. Soc. Interface* **12**: [7]

Poteete, A. R., and E. Ostrom. 2004. Heterogeneity, Group Size and Collective Action: The Role of Institutions in Forest Management. *Dev. Change* **35**:435–461. [5]

———. 2008. Fifteen Years of Empirical Research on Collective Action in Natural Resource Management: Struggling to Build Large-N Databases Based on Qualitative Research. *World Dev.* **36**:176–195. [5]

Pradelski, B. S. R., and H. P. Young. 2012. Learning Efficient Nash Equilibria in Distributed Systems. *Games Econ. Behav.* **75**:882–897. [3]

Pravosudov, V. V., and T. C. Roth. 2013. Cognitive Ecology of Food Hoarding: The Evolution of Spatial Memory and the Hippocampus. *Annu. Rev. Ecol., Evol. Syst.* **44**:173–193. [11]

Prediger, S., B. Vollan, and B. Herrmann. 2014. Resource Scarcity and Antisocial Behavior. *J. Public Econ.* **119**:1–9. [3]

Pretty, J. 2003. Social Capital and the Collective Management of Resources. *Science* **302**:1912–1914. [6]

Price, P. W. 1980. Evolutionary Biology of Parasites. Monographs in Population Biology 15. Princeton: Princeton Univ. Press. [3]

Raihani, N. J., A. Thornton, and R. Bshary. 2012. Punishment and Cooperation in Nature. *Trends Ecol. Evol.* **27**:288–295. [6]

Rainey, P. B., N. Desprat, W. W. Driscoll, and X. X. Zhang. 2014. Microbes Are Not Bound by Sociobiology: Response to Kummerli and Ross-Gillespie (2013). *Evolution* **68**:3344–3355. [7]

Rapoport, A., and A. M. Chammah. 1965. Prisoner's Dilemma: A Study in Conflict and Cooperation. Ann Arbor: Univ. of Michigan Press. [3]

Rasko, D. A., and V. Sperandio. 2010. Anti-Virulence Strategies to Combat Bacteria-Mediated Disease. *Nat. Rev. Drug Discov.* **9**:117–128. [7]

Rasolofoson, R. A., P. J. Ferraro, C. N. Jenkins, and J. P. G. Jones. 2015. Effectiveness of Community Forest Management at Reducing Deforestation in Madagascar. *Biol. Conserv.* **184**:271–277. [5]

Ratledge, C., and L. G. Dover. 2000. Iron Metabolism in Pathogenic Bacteria. *Annu. Rev. Microbiol.* **54**:881–941. [3]

Read, A. F., S. J. Baigent, C. Powers, et al. 2015. Imperfect Vaccination Can Enhance the Transmission of Highly Virulent Pathogens. *PLoS Biol.* **13**:e1002198. [9]

Read, A. F., and M. J. MacKinnon. 2008. Pathogen Evolution in a Vaccinated World. In: Evolution in Health and Disease (2nd edition), ed. S. C. Stearns and J. C. Koella, pp. 139–152. Oxford: Oxford Univ. Press. [9]

Rendell, L., L. Fogarty, and K. N. Laland. 2010. Rogers' Paradox Recast and Resolved: Population Structure and the Evolution of Social Learning Strategies. *Evolution* **64**:534–548. [3]

Repka, J., and M. R. Gross. 1995. The Evolutionarily Stable Strategy under Individual Condition and Tactic Frequency. *J. Theor. Biol.* **176**:27–31. [12]

Richerson, P. J., and R. Boyd. 2005. Not by Genes Alone: How Culture Transformed Human Evolution. Chicago: Univ. of Chicago Press. [3]

Richner, H., and P. Heeb. 1996. Communal Life: Honest Signaling and the Recruitment Center Hypothesis. *Behav. Ecol.* **7**:115–118. [11]

Rieucau, G., and L.-A. Giraldeau. 2009. Persuasive Companions Can Be Wrong: The Use of Misleading Social Information in Nutmeg Mannikin. *Behav. Ecol.* **20**:1217–1222. [11]

Robinson, B. E., M. B. Holland, and L. Naughton-Treves. 2014. Does Secure Land Tenure Save Forests? A Meta-Analysis of the Relationship between Land Tenure and Tropical Deforestation. *Global Environ. Change* **29**:281–293. [5]

Rogers, A. R. 1988. Does Biology Constrain Culture. *Am. Anthropol.* **90**:819–831. [3]

Rohwer, S., and P. W. Ewald. 1981. The Cost of Dominance and Advantage of Subordination in a Badge Signalling System. *Evolution* **35**:441–454. [4]

Ross-Gillespie, A., M. Weigert, S. P. Brown, and R. Kummerli. 2014. Gallium-Mediated Siderophore Quenching as an Evolutionarily Robust Antibacterial Treatment. *Evol. Med. Public Health* **2014**:18–29. [7]

Roth, A. E. 2015. Who Gets What – and Why: The New Economics of Matchmaking and Market Design. New York: Houghton Mifflin Harcourt. [10]

Roth, A. E., and I. Erev. 1995. Learning in Extensive-Form Games: Experimental Data and Simple Dynamic Models. *Games Econ. Behav.* **8**:164–212. [2]

Roux, D., O. Danilchanka, T. Guillard, et al. 2015. Fitness Cost of Antibiotic Susceptibility During Bacterial Infection. *Sci. Transl. Med.* **7**:297ra114. [8]

Rudel, T. K., O. T. Coomes, E. Moran, et al. 2005. Forest Transitions: Towards a Global Understanding of Land Use Change. *Global Environ. Change* **15**:23–31. [5]

Rudholm, N. 2002. Economic Implications of Antibiotic Resistance in a Global Economy. *J. Health Econ.* **21**:1071–1083. [8]

Rumbaugh, K. P., S. P. Diggle, C. M. Watters, et al. 2009. Quorum Sensing and the Social Evolution of Bacterial Virulence. *Curr. Biol.* **19**:341–345. [7]

Rutherford, S. T., and B. L. Bassler. 2012. Bacterial Quorum Sensing: Its Role in Virulence and Possibilities for Its Control. *Cold Spring Harb. Perspect. Med.* **2**:a012427. [7]

Sachs, J. D., and A. M. Warner. 2001. The Curse of Natural Resources. *Eur. Econ. Rev.* **45**:827–838. [3]

Safi, K., and G. Kerth. 2007. Natural History Miscellany: Comparative Analyses Suggest That Information Transfer Promoted Sociality in Male Bats in the Temperate Zone. *Am. Nat.* **170**:465–472. [12]

Samuelson, L. 2001. Analogies, Adaptations and Anomalies. *J. Econ. Theory* **97**:320–367. [2]

Sandoz, K. M., S. M. Mitzimberg, and M. Schuster. 2007. Social Cheating in Pseudomonas aeruginosa Quorum Sensing. *PNAS* **104**:15876–15881. [12]

Schaller, G. B. 1972. The Serengeti Lion: A Study of Predator-Prey Relations. Wildlife Behavior and Ecology, No. 86. Chicago: Univ. of Chicago Press. [3]

Schuett, W., T. Tregenza, and S. R. X. Dall. 2010. Sexual Selection and Animal Personality. *Biol. Rev. Camb. Philos. Soc.* **85**:217–246. [11]

Schuster, M., D. J. Sexton, S. P. Diggle, and E. P. Greenberg. 2013. Acyl-Homoserine Lactone Quorum Sensing: From Evolution to Application. *Annu. Rev. Microbiol.* **67**:43–63. [7]

Selten, R., and P. Hammerstein. 1984. Gaps in Harley Argument on Evolutionarily Stable Learning Rules and in the Logic of Tit for Tat. *Behav. Brain Sci.* **7**:115–116. [3]

Seppänen, J. T., J. T. Forsman, M. Mönkkönen, and R. L. Thomson. 2007. Social Information Use Is a Process across Space, Time and Ecology, Reaching Heterospecifics. *Ecology* **88**:1622–1633. [11]

Sethi, R., and E. Somanathan. 1996. The Evolution of Social Norms in Common Property Resource Use. *Am. Econ. Rev.* **86**:766–788. [6]

Shaw, R. C., and N. S. Clayton. 2013. Careful Cachers and Prying Pilferers: Eurasian Jays (*Garrulus glandarius*) Limit Auditory Information Available to Competitors. *Proc. R. Soc. Lond. B* **280**:20122238. [11]

Shettleworth, S. J. 1990. Spatial Memory in Food-Storing Birds. *Phil. Trans. R. Soc. B* **329**:143–151. [11]

Sigel, S. P., S. Lanier, V. S. Baselski, and C. D. Parker. 1980. *In Vivo* Evaluation of Pathogenicity of Clinical and Environmental Isolates of *Vibrio cholerae*. *Infect. Immun.* **28**:681–687. [9]

Silk, J. B. 1982. Altruism among Female *Macaca radiata:* Explanations and Analysis of Patterns of Grooming and Coalition-Formation. *Behaviour* **79**:162–188. [12]

Simon, H. A. 1956. Rational Choice and the Structure of the Environment. *Psychol. Rev.* **63**:129–138. [2]

Singer, R. S., R. Finch, H. C. Wegener, et al. 2003. Antibiotic Resistance: The Interplay between Antibiotic Use in Animals and Human Beings. *Lancet Infect. Dis.* **3**:47–51. [8]

Sirot, E. 2000. An Evolutionarily Stable Strategy for Aggressiveness in Feeding Groups. *Behav. Ecol.* **11**:351–356. [11]

Skaperdas, S. 1996. Contest Success Functions. *J. Econ. Theory* **7**:283–290. [12]

———. 1998. On the Formation of Alliances in Conflict and Contests. *Public Choice* **96**:25–42. [12]

Smith, J. 2001. The Social Evolution of Bacterial Pathogenesis. *Proc. R. Soc. Lond. B* **268**:61–69. [7]

Stein, G. E. 2005. Antimicrobial Resistance in the Hospital Setting: Impact, Trends, and Infection Control Measures. *Pharmacotherapy* **25**:44S–54S. [9]

Stephens, D. W. 1981. The Logic of Risk-Sensitive Foraging Preferences. *Anim. Behav.* **29**:628–629. [3]

Stephens, P. A., A. F. Russell, A. J. Young, W. J. Sutherland, and T. H. Clutton-Brock. 2005. Dispersal, Eviction, and Conflict in Meerkats (*Suricata suricatta*): An Evolutionarily Stable Strategy Model. *Am. Nat.* **165**:120–135. [12]

Sueur, C., A. J. King, L. Conradt, et al. 2011. Collective Decision-Making and Fission-Fusion Dynamics: A Conceptual Framework. *Oikos* **120**:1608–1617. [12]

Sumaila, U. R., and C. J. Walters. 2005. Intergenerational Discounting: A New Intuitive Approach. *Ecol. Econ.* **52**:135–142. [6]

Taylor, P. D., and L. B. Jonker. 1978. Evolutionarily Stable Strategies and Game Dynamics. *Math. Biosci.* **40**:145–156. [3]

Taylor, T. B., A. M. Rodrigues, A. Gardner, and A. Buckling. 2013. The Social Evolution of Dispersal with Public Goods Cooperation. *J. Evol. Biol.* **26**:2644–2653. [9]

Templeton, J. J., and L.-A. Giraldeau. 1995. Public Information Cues Affect the Scrounging Decisions of Starlings. *Anim. Behav.* **49**:1617–1626. [11]

The malERA Consultative Group on Vector Control. 2011. A Research Agenda for Malaria Eradication: Health Systems and Operational Research. *PLoS Med* **8**:e1000397. [9]

Thomas, S., M. Harding, S. C. Smith, et al. 2012. Cd24 Is an Effector of HIF-1 Driven Primary Tumor Growth and Metastasis. *Cancer Res.* **72**:5600–5612. [10]

Tisdell, C. 1992. Exploitation of Techniques That Decline in Effectiveness with Use. *Public Finance* **37**:428–437. [8]

Tóth, Z., V. Bókony, Á. Z. Lendvai, et al. 2009. Effects of Relatedness on Social-Foraging Tactic Use in House Sparrows. *Anim. Behav.* **77**:337–342. [4]

Toyokawa, W., H. R. Kim, and T. Kameda. 2014. Human Collective Intelligence under Dual Exploration-Exploitation Dilemmas. *PLoS One* **9**:9. [12]

Tracy, N. D., and J. W. Seaman. 1995. Properties of Evolutionarily Stable Learning Rules. *J. Theor. Biol.* **177**:193–198. [3]

Traxler, C. 2010. Social Norms and Conditionally Cooperative Tax Payers. *Eur. J. Polit. Econ.* **26**:89–103. [2]

Tucker, A. W. 1983. The Mathematics of Tucker: A Sampler. *Two-Year Coll. Math. J.* **14**:228–232. [3]

Tullock, G. 1967. Welfare Costs of Tariffs, Monopolies, and Theft. *Western Econ. J.* **5**:224–232. [3]

Tullock, G. 1974. The Social Dilemma: The Economics of War and Revolution. The Public Choice Society Book and Monograph Series. Blacksburg, VA: University Publications. [3]

UN. 2015. Transforming Our World: The 2030 Agenda for Sustainable Development. Secondary Transforming Our World: The 2030 Agenda for Sustainable Development. https://sustainabledevelopment.un.org/post2015/transformingourworld. (accessed Aug. 18, 2016). [5]

UNFCCC. 2015. Paris Agreement: United Nations Framework Convention on Climate Change (UNFCCC). In. https://unfccc.int/resource/docs/2015/cop21/eng/l09r01.pdf. (accessed Aug. 20, 2016). [5]

Uriarte, M., M. Pinedo-Vasquez, R. S. DeFries, et al. 2012. Depopulation of Rural Landscapes Exacerbates Fire Activity in the Western Amazon. *PNAS* **109**:21546–21550. [5]

Vale, P. F., A. Fenton, and S. P. Brown. 2014. Limiting Damage during Infection: Lessons from Infection Tolerance for Novel Therapeutics. *PLoS Biol.* **12(1)**:e1001769.

Valone, T. J. 2007. From Eavesdropping on Performance to Copying the Behavior of Others: A Review of Public Information Use. *Behav. Ecol. Sociobiol.* **62**:1–14. [11]

Valone, T. J., and L.-A. Giraldeau. 1993. Patch Estimation by Group Foragers: What Information Is Used? *Anim. Behav.* **45**:721–728. [11]

Vander Wall, S. B. 1990. Food Hoarding in Animals. Chicago: Univ. of Chicago Press. [11]

van der Weele, J. 2012a. The Signaling Power of Sanctions in Social Dilemmas. *Journal of Law, Economics and Organization* **28(1)**:103–126. [2]

———. 2012b. Beyond the State of Nature: Introducing Social Interactions in the Economic Model of Crime. *Rev. Law Econ.* **8**:401–432. [2]

Velicer, G. J., L. Kroos, and R. E. Lenski. 2000. Developmental Cheating in the Social Bacterium *Myxococcus xanthus*. *Nature* **404**:598–601. [12]

Vickery, W. L., L.-A. Giraldeau, J. J. Templeton, D. L. Kramer, and C. A. Chapman. 1991. Producers, Scroungers, and Group Foraging. *Am. Nat.* **137**:847–863. [3, 4, 12]

von Clausewitz, C. 1832/1968. On War, trans. J. J. Graham. Annotated Reprint of the Original (1832). London: Penguin Classics. [10]

von Hunolstein, C., F. Scopetti, A. Efstratiou, and K. Engler. 2002. Penicillin Tolerance Amongst Non-Toxigenic *Corynebacterium diphtheriae* Isolated from Cases of Pharyngitis. *J. Antimicrob. Chemother.* **50**:125–128. [9]

Von Neumann, J., and O. Morgenstern. 1953. Theory of Games and Economic Behavior. Wiley Science Edition. Princeton: Princeton Univ. Press. [3]

Wärneryd, K. 1998. Distributional Conflict and Jurisdictional Organization. *J. Public Econ.* **69**:435–450. [12]

Weitzman, M. 1998. Why the Far-Distant Future Should Be Discounted at Its Lowest Possible Rate. *J. Environ. Econ. Manage.* **36**:201–208. [9]

Welbergen, J. A., and N. B. Davies. 2009. Strategic Variation in Mobbing as a Front Line of Defense against Brood Parasitism. *Curr. Biol.* **19**:235–240. [12]

West, P., J. Igoe, and D. Brockington. 2006a. Parks and People: The Social Impact of Protected Areas. *Annu. Rev. Anthropol.* **35**:251–277. [5]

West, S. A., and A. Buckling. 2003. Cooperation, Virulence and Siderophore Production in Bacterial Parasites. *Proc. Roy. Soc. B* **270**:37–44. [3]

West, S. A., S. P. Diggle, A. Buckling, A. Gardner, and A. S. Griffins. 2007a. The Social Lives of Microbes. *Annu. Rev. Ecol., Evol. Syst.* **38**:53–77. [7]

West, S. A., and A. Gardner. 2013. Adaptation and Inclusive Fitness. *Curr. Biol.* **23**:R577–R584. [3]

West, S. A., A. Gardner, D. M. Shuker, et al. 2006b. Cooperation and the Scale of Competition in Humans. *Curr. Biol.* **16**:1103–1106. [12]

West, S. A., A. S. Griffin, and A. Gardner. 2007b. Evolutionary Explanations for Cooperation. *Curr. Biol.* **17**:R661–R672. [7, 12]

———. 2007c. Social Semantics: Altruism, Cooperation, Mutualism, Strong Reciprocity and Group Selection. *J. Evol. Biol.* **20**:415–432. [, 12]

WHO. 2014. Antimicrobial Resistance: Global Report on Surveillance. Geneva: World Health Organization. [8]

Wilen, J. E., and S. Msangi. 2003. Dynamics of Antibiotic Use: Ecological versus Interventionist Strategies to Manage Resistance to Antibiotics. In: Battling Resistance to Antibiotics and Pesticides: An Economic Approach, ed. R. Laxminarayan, pp. 17–41. Washington, D.C.: Resources for the Future. [8, 9]

Williams, G. C. 1966. Adaptation and Natural Selection. Princeton: Princeton Univ. Press. [3]

Williams, L. J., A. J. King, and C. Mettke-Hofmann. 2012. Colourful Characters: Head Colour Reflects Personality in a Social Bird, the Gouldian Finch, *Erythrura gouldiae*. *Anim. Behav.* **84**:159–165. [12]

Wilson, E. O. 2005. Kin Selection as the Key to Altruism: Its Rise and Fall. *Soc. Res.* **72**:159–166. [10]

Wolf, M., G. S. van Doorn, and F. J. Weissing. 2008. Evolutionary Emergence of Responsive and Unresponsive Personalties. *PNAS* **105**:15825–15830. [11]

———. 2011. On the Coevolution of Social Responsiveness and Behavioural Consistency. *Proc. R. Soc. Lond. B* **278**:440–448. [11]

Wu, G. M., and L.-A. Giraldeau. 2005. Risky Decisions: A Test of Risk Sensitivity in Socially Foraging Flocks of *Lonchura punctulata*. *Behav. Ecol.* **16**:8–14. [3]

Xavier, J. B., W. Kim, and K. R. Foster. 2011. A Molecular Mechanism That Stabilizes Cooperative Secretions in *Pseudomonas aeruginosa*. *Mol. Microbiol.* **79**:166–179. [7]

Yamagishi, T. 1986. The Provision of Sanctioning Systems as a Public Good. *J. Pers. Soc. Psychol.* **51**:110–116. [10]

Yamagishi, T., and N. Mifune. 2008. Does Shared Group Membership Promote Altruism? Fear, Greed, and Reputation. *Ration. Soc.* **20**:5–30. [12]

Young, H. P. 2008. Social Norms. Secondary Social Norms. http://www.dictionaryofeconomics.com/article?id=pde2008_S000466. (accessed Aug. 19, 2016). [6]

Zenuto, R. R., and M. S. Fanjul. 2002. Olfactory Discrimination of Individual Scents in the Subterranean Rodent *Ctenomys talarum* (Tuco-Tuco). *Ethology* **108**:629–641. [12]

Zhang, X. X., and P. B. Rainey. 2013. Exploring the Sociobiology of Pyoverdin-Producing *Pseudomonas*. *Evolution* **67**:3161–3174. [7]

Zhou, L., L. Slamti, C. Nielsen-LeRoux, D. Lereclus, and B. Raymond. 2014. The Social Biology of Quorum Sensing in a Naturalistic Host Pathogen System. *Curr. Biol.* **24**:2417–2422. [9]

Subject Index

Further Titles in the Strüngmann Forum Report Series[1]

Better Than Conscious? Decision Making, the Human Mind, and Implications For Institutions
edited by Christoph Engel and Wolf Singer, ISBN 978-0-262-19580-5

Clouds in the Perturbed Climate System: Their Relationship to Energy Balance, Atmospheric Dynamics, and Precipitation
edited by Jost Heintzenberg and Robert J. Charlson, ISBN 978-0-262-01287-4

Biological Foundations and Origin of Syntax
edited by Derek Bickerton and Eörs Szathmáry, ISBN 978-0-262-01356-7

Linkages of Sustainability
edited by Thomas E. Graedel and Ester van der Voet, ISBN 978-0-262-01358-1

Dynamic Coordination in the Brain: From Neurons to Mind
edited by Christoph von der Malsburg, William A. Phillips and Wolf Singer, ISBN 978-0-262-01471-7

Disease Eradication in the 21st Century: Implications for Global Health
edited by Stephen L. Cochi and Walter R. Dowdle, ISBN 978-0-262-01673-5

Animal Thinking: Contemporary Issues in Comparative Cognition
edited by Randolf Menzel and Julia Fischer, ISBN 978-0-262-01663-6

Cognitive Search: Evolution, Algorithms, and the Brain
edited by Peter M. Todd, Thomas T. Hills and Trevor W. Robbins, ISBN 978-0-262-01809-8

Evolution and the Mechanisms of Decision Making
edited by Peter Hammerstein and Jeffrey R. Stevens, ISBN 978-0-262-01808-1

Language, Music, and the Brain: A Mysterious Relationship
edited by Michael A. Arbib, ISBN 978-0-262-01962-0

Cultural Evolution: Society, Technology, Language, and Religion
edited by Peter J. Richerson and Morten H. Christiansen, ISBN 978-0-262-01975-0

Schizophrenia: Evolution and Synthesis
edited by Steven M. Silverstein, Bita Moghaddam and Til Wykes, ISBN 978-0-262-01962-0

Rethinking Global Land Use in an Urban Era
edited by Karen C. Seto and Anette Reenberg, ISBN 978-0-262-02690-1

Trace Metals and Infectious Diseases
edited by Jerome O. Nriagu and Eric P. Skaar, ISBN 978-0-262-02919-3

Translational Neuroscience: Toward New Therapies
edited by Karoly Nikolich and Steven E. Hyman, ISBN: 9780262029865

---

[1]   available at https://mitpress.mit.edu/books/series/str%C3%BCngmann-forum-reports-0

The Pragmatic Turn: Toward Action-Oriented Views in Cognitive Science
edited by Andreas K. Engel, Karl J. Friston and Danica Kragic
ISBN: 978-0-262-03432-6

Complexity and Evolution: Toward a New Synthesis for Economics
edited by David S. Wilson and Alan Kirman, ISBN: 9780262035385

Computational Psychiatry: New Perspectives on Mental Illness
edited by A. David Redish and Joshua A. Gordon, ISBN: 9780262035422